ASSISTING THE INVISIBLE HAND

Issues in Business Ethics

VOLUME 18

The titles published in this series are listed at the end of this volume.

Assisting the Invisible Hand

Contested Relations Between Market, State and Civil Society

by

WIM DUBBINK

Faculty of Social Sciences,
Vrije Universiteit Amsterdam, The Netherlands

KLUWER ACADEMIC PUBLISHERS
DORDRECHT / BOSTON / LONDON

A C.I.P. Catalogue record for this book is available from the Library of Congress.

ISBN 1-4020-1444-9

Published by Kluwer Academic Publishers,
P.O. Box 17, 3300 AA Dordrecht, The Netherlands.

Sold and distributed in North, Central and South America
by Kluwer Academic Publishers,
101 Philip Drive, Norwell, MA 02061, U.S.A.

In all other countries, sold and distributed
by Kluwer Academic Publishers,
P.O. Box 322, 3300 AH Dordrecht, The Netherlands.

The Netherlands Organization for Scientific Research (Nederlandse Organisatie voor
Wetenschappelijk Onderzoek) has financially supported the publication of this book. English
translation: Taalcentrum-VU, Amsterdam, The Netherlands, David Doherty.

Printed on acid-free paper

For my daughter
And in memory of my mother

'A people among whom there is no habit of spontaneous action for a collective interest - who look habitually to their government to command or prompt them in all matters of joint concern - who expect to have everything done for them … have their faculties only half developed: their education is defective in one of its most important branches. …'

'There cannot be a combination of circumstances more dangerous to human welfare, than that in which intelligence and talent are maintained at a high standard within a governing corporation, but starved and discouraged outside the pale. Such a system, more completely than any other, embodies the idea of despotism …'

'It is therefore of supreme importance that all classes of the community, down to the lowest, should have much to do for themselves; that as great a demand should be made upon their intelligence and virtue as it is in any respect equal to; that the government should not only leave as far as possible to their own faculties the conduct of whatever concerns them alone, but should suffer them, or rather encourage them, to manage as many as possible of their joint concerns by voluntary co-operation.'

J. S. Mill
Principles of Political Economy (1848) 308-309

TABLE OF CONTENTS

PREFACE

Wim Dubbink's research spans three different levels, which shift in and out of focus in the course of his book. First of all there is the *global question* which addresses the complex relationship between market, state and civil society. While this question lies at the heart of his book, there is far more to it than that. This relationship is examined in relation to the *three disciplines* which are concerned with these fundamental institutions: economics, public administration and political philosophy. *Assisting the Invisible Hand* is an exploration of the way in which these three academic disciplines interpret the mutual relationships between the three fundamental institutions.

Even this does not fully reflect the scope of the work. The three fundamental institutions and the disciplines they have spawned are not examined for theoretical reasons, however important the theoretical content of this book may be. The problems facing the *environment* form the inspiration for the author's inquiry. It is from this field that he plucks his crucial examples and this brings us to the third and by no means least important level on which the book operates. The 'invisible hand' needs assistance when it comes to the question of if and how *sustainability* can be realized in complex societies like ours.

In this preface I would like to address each of these three levels.

1. The global question would appear to be self-evident but, of course, this is far from being the case. All manner of political movements have, until recently, focused on a single fundamental institution as the model on which their world view is based. Laissez-faire liberalism was an ideology of the market. Socialism, both in its social-democratic and its totalitarian manifestations, was primarily focused on the state as the source of the redistribution of wealth. As these movements stood tall, civil society remained in the shadows, only to undergo a re-evaluation in the 1960s and 1970s fuelled by the rise of the student movement and various forms of cultural anarchism. As one might expect from such a history, there is a great deal of ideological rubble to be cleared away before the mutual relations between the three institutional spheres can be exposed and illuminated without resorting to oversimplification.

2. This book clears the way for this task in two ways. To begin with, the author, as I have said, allocates a central position to the three disciplines within which the fundamental institutions are addressed. *Assisting the Invisible Hand* belongs from this point of view to the field of scientific study. The institutions with which this book is concerned are not to be found in society in any kind of pure form. There is no one who can simply point the way to civil society, the state or even the market in the broadest sense of the term. At the risk of sounding a tad too trendy, it might be said that the market, the state and civil society are in fact constructs thought up by philosophers, academics and politicians and sustained subsequently in the various arenas of ongoing debate.

Secondly, this book looks at the way in which the three disciplines each consider the mutual relationship between the three fundamental institutions from their own perspective. The author does this by identifying the 'mental models of social organization' which play a leading role in these disciplines. He uses this term to refer to the image, the implicit conception each discipline maintains with regard to the three spheres and it is precisely on this point that I find his research so fascinating and groundbreaking. The study embarks upon a painstaking archaeological dig to unearth these underlying concepts. For example, the neoclassical economists are shown to operate according to concepts which implicitly determine our relationship with both the market and the state. Without the restrictive conditions placed by the state upon the market, the market is revealed as not being able to function. It is barely possible to imagine a greater difference between this and laissez-faire liberalism, which ideologically speaking is still very much alive today.

It is these often partially implicit mental models which the author is eager to pursue. He divests them of their academic trappings, lays them on the operating table and picks them apart with anatomical precision. It is ultimately their shortcomings that enable the author to present his own model for the relations between state, market and civil society in the closing chapter.

A society that holds sustainability in high esteem would appear not to be able to do without any one of the three fundamental institutions. This makes all the more pressing the question of how these institutions relate to one another and what barriers ought to exist between them. Dubbink leaves the reader repeatedly astounded at how easily all manner of scientific theories about society as a whole simply fail to address key aspects of this issue. Is this failing due to the academic tendency towards specialization? Or are we simply incapable of arriving at new insights into disciplines of this kind without the threat of urgent global problems like environmental crises or famine? When I think of the work of pioneers like Pigou and Sen, I feel compelled to answer this latter question in the affirmative.

3. Whatever the case might be, in Dubbink's view state, market and civil society form three essential and non-interchangeable perspectives for the analysis of environmental problems. This has far-reaching consequences, two of which I would now like to mention. The first is *methodological* in nature. Here society is viewed in an institutional light. The book's ironic title already suggests as much. After all, since Smith the 'invisible hand' has been the metaphor for a process that has been assumed to take place behind the backs, as it were, of the social actors. Accordingly it was a process which by definition did not require assistance, an idea that was adopted from Smith by Hegel and Marx. True, the actions of individuals are an essential part of the market process, but their institutional consequences are not the result of intention. These consequences are beyond the short-term interests of actors and groups. The invisible hand may indeed need some help, but it is not under threat of elimination. A theory of sustainability which thinks it can manage without the institutional power of the market process, positions itself outside of the social reality. The market alone cannot solve the problems facing the environment, but without the market success is just as unlikely. Whatever measures need to be taken, in one way or another they will have to be compatible with a relatively autonomous market in which the pursuit of profit is institutionalized.

This book is characterized by a high level of social realism and, following in the footsteps of Cohen and Arato, the associated criticism of utopian notions. The solutions sought after by Dubbink are particularly geared towards institutional change and are continually concerned with the boundaries between the market, state and civil society, which he regards as movable and open to discussion. He does not try to wriggle out of problems, for example by confining himself to an appeal to the individual consumer and all manner of positive alterations in his or her behaviour. The changes necessitated by present-day environmental dilemmas are too fundamental to be realized by means of individual good intentions.

As well as having a methodological effect, the required symbiosis of state, market and civil society also has *political* consequences. The only political system which can provide an anchor for all three fundamental institutions, and which also exists by the grace of this combined approach, is *liberal democracy*. Accordingly, Dubbink devotes a great deal of attention to this very system. His analysis of the role of the state takes place within this context. Despite all kinds of predictions as to its imminent downfall, the state still embodies an enormous power which in the interests of democracy also makes a powerful civil society desirable. The opposite is equally true: a strong civil society cannot do without the indirect channel of a government and its attendant systems of representation. The author also gives short shrift to theories that base themselves on one form of direct democracy or another, because of the high degree of utopianism they exhibit. These, too, are found to be lacking that most vital of ingredients: social realism.

The reader of this introduction will doubtless want to know which models of social organization the author draws from the disciplines he has studied and which proposals he goes on to present himself. I do not regard it as part of my remit to make such revelations. A good preface should not attempt to stand in for a book but should stand aside and make way for it. The reader will discover along the way that this author is a frontiersman and a fascinating guide for all who share his interest in the exploration of borders. The path he sets out is a clear one and those who accompany him need never fear losing their bearings. His arguments are erudite and there is much to be learned along the way. The resulting work represents an important contribution towards solving a problem that remains underestimated. In other words it is a journey that will not be taken in vain.

Lolle Nauta
Professor Emeritus of Social Philosophy
University of Groningen
The Netherlands

ACKNOWLEDGEMENTS

This book is a thoroughly revised edition of the PhD thesis that I wrote a few years ago under the supervision of Prof. L.W. Nauta and Prof. J. Kooiman. Circumstances forced me to complete some parts of my thesis rather hurriedly. Later on I grew dissatisfied with the end result. I am therefore extremely grateful to Prof. H. van Luijk for giving me the opportunity to rewrite and adapt the text for an international audience. There is another important reason why I am indebted to Prof. Van Luijk. A peculiar circumstance that hampered me during the writing of my original thesis was the lack of a scientific discourse which dealt with the same questions that fascinated me. I was writing about economists, administrative theorists and political philosophers but their research questions were always some way removed from mine. Prof. Van Luijk explained to me that the questions I was asking were closely connected to theoretical discussions within contemporary business ethics. And indeed, when I looked at my research from the perspective of business ethics it became far easier for me to formulate the central theme of my research. Accordingly I am glad that the book is now being published as part of the series *Issues in Business Ethics*.

Prof. L.W. Nauta was kind enough to give a critical evaluation of the new manuscript and also wrote the preface. I am very grateful for his valuable contributions. I would also like to thank the Netherlands Organization for Scientific Research for the grant that covered the translation. David Doherty, employed by the translation department of Taalcentrum-VU in Amsterdam, was responsible for the translation. I think he did an excellent job. Finally, I wish to thank the reviewers (whose names are not known to me) for their useful suggestions.

CHAPTER 1

INTRODUCTION

1.1 PUBLIC RESPONSIBILITY IN THE MARKET

Businesses have a role to play in improving the lives of all their customers, employees, and shareholders by sharing with them the wealth they have created. Suppliers and competitors as well should expect businesses to honor their obligations in a spirit of honesty and fairness. As responsible citizens of the local, national, regional and global communities in which they operate, businesses share a part in shaping the future of those communities

A business should protect and, where possible, improve the environment, promote sustainable development, and prevent the wasteful use of natural resources.

Caux Round Table, Principles for Business
Section 2, general principles, Principle 1 (second section) and Principle 6
www://cauxroundtable.org

The free market is often characterized as a social domain within which people act in accordance with their own self-interest (Habermas, 1981; Alec Gee, 1991: 105-106. See also: Boatright, 1999). According to this widely held view, actors in the market only have a small and restricted responsibility for the public problems society has to cope with, even if those problems originate in market processes. From this perspective, the market can therefore be referred to as the sphere of limited public responsibility.

The proposition that the market is the sphere of limited responsibility is interpreted by the advocates of the position described above as both an empirical statement and as a normative statement. They claim that, empirically, self-interest is by far the main motivation of actors in the market. At the same time, they maintain that this is exactly how the market is supposed to operate. An actor should not be morally blamed for concentrating on his own interests in the market. Advocates of this view differ as to the reasons behind their normative claim. Some believe that the market fulfils its societal functions best if actors concentrate on their self-interest. Others hold that the market defends important values, such as the freedom of the individual, most effectively in this way.

Those who adhere to the limited responsibility view generally acknowledge that modern Western society is troubled by grave public issues. They also admit that the causes of some of these public issues originate in the market. Environmental problems are a case in point. However, they hold that if the market has to be controlled in order to alleviate some public problem, this must be done by changing the laws and other limiting conditions within which the market process is embedded.

1

This view has four implications for thinking on controlling the market. Firstly, within this view controlling the market with regard to public issues becomes *exclusively* a matter of tinkering with limiting conditions. Since these limiting conditions are always institutions, Kettner (1994: 247) has characterized this way of thinking on controlling the market as *institutionalism*. Secondly, this way of thinking makes a strong appeal to the state, because the state is the only institution within modern liberal-democratic society that is in a position to change the limiting conditions of the market. Thirdly, this view tries to control the market at *system level*. Fourthly, it presupposes a sharp division of labour between the tasks of the state and the tasks that are to be fulfilled in the market. I will henceforth refer to this way of thinking about controlling the market as the *'indirect responsibility model'*.

As a way of thinking about controlling the market, the indirect responsibility model has been dominant within Western liberal democracies throughout the 20th century. Nevertheless, its dominance was never so great that it completely did away with other views. In recent decades the dominant position of the indirect responsibility model has been undermined. One of the factors in this process has been the rise of all sorts of theories which stress the problems of state action (Pressman and Wildavski, 1973; Huntingdon, 1975; Buchanan, 1977; Bobbio, 1984; Held, 1983 and 1987 Mayntz, 1978; Yeager, 1991). As a consequence one alternative model of controlling the market has gained strength. I will refer to this alternative model as the *'direct responsibility model'*.

The direct responsibility model differs remarkably from the indirect responsibility model in its view on controlling the market. The core of the model is that actors on the market should personally take some responsibility for dealing with public issues. This model has a number of implications for thinking on controlling the market. According to the direct responsibility model, the market should not be controlled exclusively by limiting its conditions. The direct responsibility model therefore makes less of an appeal to the state. It also abandons the idea of controlling the market at system level only. Since actors are called upon to personally acknowledge their responsibility, this model places some public responsibility at *actor level*. It follows that this model also does away with the strict division of labour that characterizes the indirect responsibility model. (A schematic overview of the main aspects of both the direct responsibility model and the indirect responsibility model can be found in the appendix).

The basic question posed by this study is which model of controlling the market is most plausible and normatively preferable. It is important to realize, however, that the two models are not diametrically opposed to one another. The indirect responsibility model is not the opposite of the direct responsibility model. It does not absolve people of their responsibilities. It is not a plea for anarchy, indifference or egoism. It merely states that in modern society citizens can and ought to transfer their responsibilities for public issues to the state. On the other hand, the direct responsibility model does not deny that limiting the conditions of the market is an important way to control it. Nor does the direct responsibility deny that the state has an important role to play in controlling the market. It only objects to the exclusive role that the indirect responsibility model attaches to limiting conditions and the role of the state. That is why it would be wrong to typify the direct responsibility model as 'voluntarism' as opposed to the institutionalism of the indirect responsibility model. If 'voluntarism' means that one conceptualizes the problem of controlling the market entirely at actor level by trying to normatively adjust the conduct of individuals, then the direct responsibility model is only *partly* voluntarist. It has an institutionalist component as well. The indirect responsibility model, however, is characterized by its *extreme* institutionalism.

The two models are also not diametrically opposed with regard to their views of the general relation between morality and the market. Advocates of both models explicitly distance themselves from the notion that the market can be regarded as the domain in which it is the norm for people to act immorally or amorally, regardless of whether one interprets this proposition empirically or normatively. They both wholeheartedly endorse the proposition that actors on the market have to behave morally, in the sense that they are at least expected to abide by the law and the basic rules of decency. If people transgress these standards on a regular basis, they undermine the desirability of the market as an institution. It would also no longer be possible for either the state or the market to function adequately.

Another important moral duty about which the advocates of both models agree is that market parties ought to act with integrity in the political arena. Baumol (1975) calls this the meta-duty of market parties; Ulrich (1997: 434) calls it the republican duty. The republican duty implies important prohibitions. Companies, for example, are not allowed to use their position of power to frustrate the drawing up of laws and regulations (e.g. by making use of information asymmetries). The republican duty also contains an important command. Companies have to inform the citizens and the state of issues in their sector which are in need of regulation from a public viewpoint.

One can therefore conclude that, with regard to the general relation between morality and the market, both models place a fair moral burden on market actors (see also: Bowie: 1989). But they differ when it comes to controlling the market in order to deal with public issues. The indirect responsibility model holds that the responsibility of actors in the market ends with their republican duty, abiding the law and the rules of common decency. The direct responsibility model calls on market actors to use at least some of their freedom (discretionary space) to solve public problems. In other words, the indirect responsibility model by and large maintains the idea that people in the market take on the role of economic man. The direct responsibility model involves a drastic revision of this view: in the market, man also has to operate as a citizen, at least to some extent.

Mental models

Theoretical discussions within contemporary political theory often deal with various interpretations of central liberal-democratic values such as freedom or equality. In the light of this preoccupation, the present study is unconventional in the level of its analysis. It focuses on two different ways of thinking about the *organization* of the social order. I refer to these ways of thinking as 'mental models of social organization'. A mental model of social organization is a model which indicates what the relationship between the state, the free market and civil society should be like and which also addresses the main structural aspects of each of these fundamental institutions. I define the concept of 'mental model' in a normatively neutral sense as a conceptual framework that people need in order to understand the institutional order (see also Werhane, 2000).

It is important to note that the debate on the adequacy of mental models is always treated as a poor relation within thinking on controlling the market. This proposition holds for all relevant academic disciplines, from economics to political theory and from business ethics to administrative theory. As a consequence the discussion on the adequacy of the mental models tends to be conducted implicitly. That is why I speak of mental *models*, as opposed to mental *theories*. A theory explicates and structures the knowledge, views and opinions that we routinely and unreflectively use in daily practice. Thinking on social organization is almost never made that explicit. It is characterized by its vagueness and the implicit ways in which it is used.

Mental models in historical perspective

If we evaluate the controversy between the indirect responsibility model and the direct responsibility model from a historical perspective, the beginning of the 21st century constitutes an interesting period. About 150 years ago laissez-faire thinking was the dominant mental model of social organization. Laissez-faire thinking shared with the indirect responsibility model the idea that actors should concentrate on their self-interest while acting on the market. It deviated from the indirect responsibility model in its tendency to downplay the role of the state in relation to public issues. According to laissez-faire thinking public problems were inevitable and always hard to deal with, irrespective of the means chosen. However, trying to deal with them by means of the state was probably considered to be one of the worst options. The general view was that state action was likely to end in disaster.

If we look back on laissez-faire thinking with the benefit of hindsight we may conclude that it was too pessimistic about the potential of the state and overly optimistic about the ways in which public issues could be controlled spontaneously by market forces. However, any assessment of laissez-faire thinking must bear in mind that historically it was a critical theory which reacted against the status quo most of the time. Mannheim's (1950) distinction between an 'ideology' and a 'utopia' is relevant in this context. Mannheim defined an 'ideology' as a model that sustains the social and political status quo. Ideologies are always quite elaborate. A 'utopia' is a model that is used by its proponents as a means of bringing about social or political change. Its primary function is critical. That is why the model often has blanks and underdeveloped aspects. (However, this is not to say that ideologies are always more adequate than utopias. Mannheim stated that since ideologies sustain the status quo, they stop short of confronting present reality.)

Laissez-faire thinking has long been a utopia in the Mannheimian sense. Its finest hour was the relatively short period between 1830 and 1870. In that period the free market was present in its most unfettered form in some Western societies, notably Great Britain (Polanyi, 1944; Searle, 1998; Gray, 1998). This caused increasing numbers of people to develop at the very least an ambivalent attitude towards it, partly because of the unintended but grave effects which market processes had on public concerns such as poverty or the environment. Many liberal-democratic thinkers like Dewey (1905; 1932) Pigou (1920) Keynes (1926) and Wicksell (1934) recommended that the market should be controlled somehow in order to make its outcome more desirable. With the rise of Marxism and early social democracy there even came a radical rejection of the idea of the free market.

In the century after 1870 two alternative mental models of social organization took shape: the indirect responsibility model and the direct responsibility model. Both views on how to control the free market have had their champions. For example, the desire to introduce public responsibility to the market at actor level is expressed in Christian thinking on the free market (Pesch, 1920; Dooyeweerd, 1935, 1960; Banning, 1960; see also Skillen, 1974). This ideal has also inspired some socialists (Mannheim, 1951), and a fair number of liberal democrats embrace the idea as well (Dewey, 1905a and 1932i; Keynes, 1926). However, the view that the market ought to be controlled by limiting its conditions has dominated the thinking of 20th-century Western liberal democracies, both within the academic world and beyond. The reason why this approach won the day is obvious. The notion that the free market can be regulated using limiting conditions makes it possible for the state to intervene in areas where the free market leads to disadvantages, while leaving intact the ideal of a sphere where actors are free to concentrate on their economic interests. It constitutes a wonderful synthesis of the arguments for and against the free market.

Mental models and thinking on corporate social responsibility.

The indirect responsibility model is no longer taken for granted nowadays. Many academics question it from an empirical point of view. They point out that the state is a 'limited use institution' which cannot fulfil all the tasks that the proponents of the indirect responsibility model would like it to perform. Other academics have serious doubts about the indirect responsibility model from a normative perspective. They fear, for example, that the logic of the indirect responsibility model will lead to excessive interference in the market in some situations. From all sides, therefore, a need is being expressed for a mental model of social organization which is less dependent on the state to handle its affairs.

The corresponding rise of the direct responsibility model is also making itself felt in all quarters. There is, for example, a strong movement within administrative theory which advocates the direct responsibility model (Kooiman, 1993; Jentoft, 1989). Business ethics is also important in this regard (Donaldson, 1982; Van Luijk, 1994; Paine, 1997). It should be noted, however, that although much thinking in business ethics and administrative theory presupposes the direct responsibility model and as such stimulates it, there is not much theory that explicitly takes the mental models of social organization as its unit of analysis. The heart of contemporary thinking on corporate social responsibility (CSR), for example, lies with delimiting and determining the duties and responsibilities of corporations within present-day society. Questions about the necessity and the justification of the type of market that CSR presupposes are hardly ever taken up seriously. The same holds for the possible drawbacks of CSR from a liberal-democratic perspective and for the implications of CSR with regard to the relations between market, state and civil society.

Institutional order and mental models

The level of analysis of this study is the level of mental models of social organization. One can therefore look upon this research as a complement to much contemporary thinking on controlling the market, such as theories on CSR. In this analysis the indirect responsibility model and the direct responsibility model will be analysed purely as mental models of social organization. I will not look at the way in which liberal-democratic societies actually try to control the market. It is important to stress this because oftentimes researchers claim to be analysing mental models when they are actually dealing with practice (and vice versa). The problem with this mixing of levels is that there is often a considerable gap between the dominant mental model of a certain society and the way that society is actually organized. For example, Dutch administrative practice is permeated with corporatist tendencies. Yet it is hard to account for these tendencies at the level of the mental models which Dutch administrators officially embrace when they explicitly touch on the subject of market control in a liberal-democratic context.

If there are sometimes considerable discrepancies between practice and mental models, then what is the use of examining these mental models? I will try to answer this question briefly by looking into the function of mental models in social life. Social life is grounded in institutions, which together form an institutional order. Mental models are needed in order to understand and reflect upon the institutional order. I define the 'institutional order' as an entire set of (formal) rules, norms, institutions, associations and artefacts that go to constitute, facilitate, limit and perpetuate actions within a society. I echo Durkheim (see Scott, 1995: 10) in calling the institutional order the domain of social facts. Social facts present themselves to an actor as objective facts, outside of himself. However, we must be careful not to reduce the institutional order - or social facts in general - to things outside an actor. The insights of the American pragmatist John Dewey are most valuable here. Dewey (1925-1927g: 242) makes it clear that the way in which a human order exists is ontologically different to the way in which a forest, for example, exists. A forest consists of a collection of trees which need no understanding of their mutual relations to stand side by side. In that sense trees exist independently of one another. The same does not apply to a collection of people. The human order is a relational order. The relationships between people are determined by rights, obligations, rules of conduct and so on. In order to deal with this situation, people have to understand the social facts. For the stability of society in the long term it is also necessary that people see the order created by the entire set of social facts as legitimate (see also Weber, 1921: 16).

Dewey's analysis shows that the institutional order always exists on two levels simultaneously. At one level the institutional order manifests itself as an *objective* order. It is the entire structure of social facts that, to a greater or lesser extent, forces people empirically to structure their actions in specific ways. The police officer's truncheon, pistol and report book are all examples of this aspect of the institutional order. In addition to this, an institutional order also has an *interpretative* component. The institutional order is something that is understood and legitimized. The police uniform symbolizes this interpretative component. The uniform worn exclusively by the police distinguishes the police officer from an armed criminal or a village idiot who might put a stretch of red-and-white tape across the road and order people not to go any further. I refer to the interpretative component of the institutional order as the mental model of social organization. Based on the above, we can conclude that mental models have two functions for the institutional order. On the one hand they constitute the framework that makes it possible to understand the institutional order. On the other hand they make it possible to reflect upon that order, normatively and otherwise. So the study of mental models is useful even if it is not always of direct use in practice. For as soon one reflects upon the institutional order, one has to fall back on mental models.

Explanatory notes on the key questions

This analysis of the indirect responsibility model and the direct responsibility model has certain limitations and points of special interest. It is not my intention to provide a full analysis of both models. First of all, I have chosen to emphasize the normative aspects of both models. Throughout the book I will assume that modern Western societies believe they should be liberal democracies and that this is what they actually are, at least to some extent. An important part of my research will be to assess if and to what degree each model fits into the liberal-democratic tradition. In doing so, I define 'liberal democracy' as a normative theory in which the individual takes centre stage. The central values of the liberal-democratic tradition are freedom, equality, autonomy, justice and solidarity. I have added sustainability to this standard roll call of ideals. While this addition has its opponents, it is not regarded as particularly exceptional nowadays. The notion that liberal democracy implies sustainability is defended by the likes of Robert Goodin (1992). Some examples of authors I consider to be part of the liberal-democratic tradition are Buchanan, 1977; Cohen and Arato, 1992, Dahl, 1989; Dworkin, 1978; Hayek, 1976; Held, 1986; Nozick, 1974; Rawls, 1972 and Walzer, 1983. The main reason for this enumeration is to make clear it that I interpret the concept 'liberal-democratic' in a broad sense. I am not using it to denote a specific movement within the political landscape of the United States. That movement is only one manifestation of liberal democracy.

Another limitation placed on this study is its focus on two particular questions. Firstly, is there enough support for the argument that a new mental model of social organization is needed? In other words, can it actually be demonstrated that the indirect responsibility model is no longer feasible? Secondly, what might the content of the direct responsibility model be? What form does a mental model take which has at its core the idea that market actors have to bear public responsibility? As I will explain briefly, these questions seem to be the most interesting ones in the modern context.

The indirect responsibility model is due a large amount of credit. First of all, for normative reasons. As a mental model of social organization, it fits in exceptionally well with the values of liberal democracy. Civil liberties and personal freedom are important values within liberal-democratic thinking. The idea of a sphere of limited responsibility well-suits the realization of these values. Democracy is another much cherished liberal-democratic value. The notion that all public issues are controlled by the state and that the state itself is democratically governed, clearly reflects this value (see also Chapter 3). There are also practical considerations which strengthen the case in favour of the indirect responsibility model. Talking about sweeping changes to the institutional order is one thing, but realizing them is an entirely different matter. The reasons for changing a mental model of social organization therefore need to be utterly convincing.

This study has been written with the idea that many present-day arguments in favour of a new mental model of social organization fail to convince in this sense. Critics dismiss the indirect responsibility model far too easily. In particular, they underestimate its normative value. In criticizing the state's mediocre performance record, for example, they take too little account of the fact that the state in a liberal democracy is supposed to be a fettered giant. This situation would not be much different in any other liberal-democratic mental model of social organization.

Furthermore, the critics' arguments often falter in terms of their logic. They set about trying to persuade us of the dysfunction of the indirect responsibility model by detailing the state's inability to cope with the many seemingly insoluble public problems facing present-day liberal democracies, such as the problems facing the environment (Nelissen, 1992). This argument fails to convince, however. Supporters of the indirect responsibility model might reply that there are indeed many virtually insoluble public problems. They may even concede that the state does not function adequately. Yet all these admissions do not force them to admit that the indirect responsibility model is ripe for replacement. They are free to dismiss these shortcomings as implementation problems, and put them down to poor management. They can always claim that it is not the indirect responsibility model that is failing, but the people who are supposed to design and implement it.

So, in order for an analysis to show that it is time for the indirect responsibility model to step aside, it has to demonstrate that there are *structural* or *inherent problems* with this mental model in the modern-day context. Democracy can sometimes lead to fascism, as it did in Germany in 1933, but this is not sufficient reason to reject democracy as such. Democracy's descent into fascism may have been due to the specific way in which democracy was structured and operated at the time. Adapting or even reinforcing democracy in such a case might be a better way of preventing fascism than searching for a new mental model. The first question of this study is rooted in these considerations. This study contains a critical examination of the arguments against the indirect responsibility model. Do they indeed lay bare structural or inherent shortcomings?

The notion that market actors should take a certain responsibility for public issues is seen here as the core of the new direct responsibility model. But what should the rest of the new mental model of social organization look like? A mental model describes the relationship between state, market and civil society and also addresses the main structural aspects of each of these fundamental institutions. As we will see, the dominant indirect responsibility model can be reconstructed quite elaborately. This is hardly surprising since it is an ideology in Mannheim's sense of that concept. From this perspective, the direct responsibility model can best be looked upon as a utopia. All manner of normative choices still have to be made to give it more substance. For example, one could argue that the public responsibility of market parties implies that institutional structures should be set up in the market to facilitate cooperation between market parties and the state. But one might just as easily argue for a strict separation between government responsibility and that of the market parties. This study will attempt to provide greater clarity regarding the possible choices, thereby allowing the contours of the new mental model of social organization to take shape.

In performing this task, I will concentrate primarily on the normative aspects of this new mental model. The most important reason for this focus is that the normative perspective remains underexposed in the work of many other academics. For example, little attention has been paid to the problem that giving public responsibility to market parties represents a possible erosion of democracy. There are also quite a few administrative theorists whose interpretation of public responsibility entails stimulating cooperation between market parties and the state (Jentoft, 1989, 1995, 1998 and 2001; Kooiman, 1993b, 1993c and 1993d). However, from a liberal-democratic perspective cooperative ventures of this type must be regarded as highly suspect, for fear of creating a government that is too powerful and for fear that all social power will come together in one party. *'Perhaps the deepest reason for distrust in Western Europe, and particularly in the United States, of either the Russian* (communist - wd) *or the Fascist system is the unwillingness to be subject to* absolute control of a single master', wrote Dewey in *Toward the Future* (1932j: 428, stressed passage original text). A new mental model will therefore have to take into account these and many other liberal-democratic sensitivities. Mapping out further details of the new mental model will be no easy task, especially not when one considers how well the indirect responsibility model fits in with liberal-democratic thought.

Operationalizing the key questions

Karl Polanyi (1944) coined the term *'tacit dimensions of knowledge'*. Hirschman (1977: 69) defined this succinctly as *'...propositions and opinions shared by a group and so obvious to it that they are never fully or systematically articulated'*. He observed that the academic study of tacit knowledge is a tricky business. Since background knowledge remains implicit and is not systematized, studying it necessitates gathering small clues and little pieces of evidence from all over the place. To a large extent, mental models of social organization are background knowledge. The models that people use remain implicit most of the time. As a consequence, they are also lacking in detail. A study comparing two mental models of social organization therefore calls for creative operationalization. This I have attempted to do by not rushing headlong into the task of answering the key questions. Instead we will take something of a diversion, concentrating for a good while on the question of how present-day Western society thinks about the control of the free market. By studying these ways of thinking I expect that we will gradually gain an insight into the details of the mental models of social organization concealed within them.

Before embarking on this process I will first set the boundaries of the playing field in various ways. First of all, it seems wise to restrict the question of how to control the free market to one specific subject. This will allow the analysis to gain depth. Our subject has to fulfil several conditions. To begin with, it has to be a public problem. Unfortunately, this leaves us with plenty of candidates, like poverty, discrimination, public safety and so on. But the subject also has to have close links with the workings of the market, since the advocates of the direct responsibility model do not burden market parties with responsibility for public problems in general. Market parties ought to take responsibility for issues that are related to their own business. This condition makes subjects like discrimination and public safety less appropriate. Although we may encounter these problems in the market, we also encounter them in civil society. Indeed, this may even be where their roots lie. In addition to this, the subject should give clear insights into the extent and causes of the limits of state action. On the one hand this implies that the state should have considerable difficulty in coping with the problem. On the other hand it implies that our subject should be fairly well documented. This is not an empirical study after all. We are investigating the soundness of arguments for and against the two mental models of social organization.

In my opinion the issue of the environment satisfies all these conditions. The search for sustainability is a grave public issue facing modern Western societies. Environmental problems are directly linked to the workings of the free market and in recent decades the state has experienced great difficulties in coping with them. What is more, the sociological, institutional and political-administrative dimensions of environmental issues are well-documented. Many case studies and general theories have been written on the subject. Interestingly, in many of these studies the researchers trace the difficulties in solving environmental problems to the fact that environmental problems challenge the institutional order (see for example Opschoor, 1989; Freeman et al., 1973). As these researchers see it, modern society will only succeed in overcoming environmental problems if adjustments are made to the design of the institutional order. The current discussions on environmental problems therefore seem a perfect stepping stone for our research on the two mental models. In saying this, I am not implying that environmental problems are unique, but simply that environmental concerns provide a well-documented example with which to analyse the typical structure of present-day public problems.

A second limitation on the scope of this study is my decision to concentrate on present-day *academic thinking* about controlling the free market. Compared to most writers, academics are more concerned about explaining their assumptions and so this provides the greatest chance of success when it comes to uncovering the exact nature of the mental models of social organization. I will further limit the subject of my research by concentrating on three academic disciplines: *economics, administration* and *political philosophy*. Each of these academic disciplines maintains a special relationship with one of the three fundamental institutions in modern society: market, state and civil society. What is more, these three disciplines are particularly focused on sustainability as an issue of governance. However, I will not attempt to draw strict dividing lines in this respect. Many interesting authors, such as C.B. Lindblom, C.D. Stone, R. Mashaw, G. Teubner, I. Maus and A.O. Hirschman are difficult to categorize and would therefore fall outside my research field if I were to be too stringent in cordoning off these disciplines. I justify this informal approach to methodology with the argument that my research in the various disciplines is not a goal in itself. This analysis is only there to serve the systematic key questions. One last limitation built into this study is that, within each discipline, I will focus on one main stream of thought. Within economics it will be neoclassical theory, in administration it will be the theory on the limits of state action and within political philosophy it will be the emergent theory of civil society.

To summarize in terms of operationalization: in this study we are going to investigate how control of the free market is viewed in present-day neoclassical theory, in present-day administrative theory on the limits of state action and in political-philosophical theory on civil society. We will concentrate primarily on theories relating to environmental problems. The description and analysis of these theories should supply enough material to answer key questions regarding the supposed bankruptcy of the indirect responsibility model and the nature of the new mental model of social organization.

Plan of action

The structure of the study is relatively straightforward. In each of the next three chapters I will closely examine the thinking on controlling the free market in one of the academic disciplines. I will discuss each theory fairly independently. In the last section of each chapter the relationship to the key questions of this study will be established. In Chapters 2 and 3 the main focus will be on the first key question. Can it be established that the indirect responsibility model is outdated? The last two chapters are mainly devoted to the second key question about the contours of the new direct responsibility model. There I will evaluate a significant attempt to sketch the contours of the new mental model of social organization and lastly I will make some suggestions of my own. I will devote the coming sections of the present chapter to the explication of a number of important assumptions surrounding sustainability and liberal democracy.

1.2 SUSTAINABILITY AS AN EXEMPLARY PROBLEM

During one of our many car journeys through the landscape of the east of the Netherlands, my mother observed that the meadows had looked very different when she was a girl. The farmers' efforts to create a smooth grassy surface used to constantly be thwarted by daisies and other weeds. Birds like larks, godwits and pipits inhabited the pastures, together with uninvited guests like moles, mice and hares. The landscape was interspersed with small hills, slopes, ditches and wooded banks, with here and there a pool or a majestic solitary tree. Today's meadows bear little resemblance to their predecessors. They are mostly featureless green plains where grass and cattle are the only survivors. The larks have all but disappeared, the daisies have been banished. The majority of the trees have been felled, the wooded banks have been cleared and the pools filled in. Ditches have been straightened out and deepened. Even the fieldmice have fallen on hard times.

The transformation of the Dutch landscape provides a telling example of processes that have taken place at national and international level, and which are customarily grouped together under the heading 'environmental problems'. This concept covers a whole range of different phenomena, from the greenhouse effect to soil erosion and from acid rain to groundwater depletion. Environmental problems occupy a prominent place in this study. Our present inquiry into the regulation of the free market centres around the example of this public issue. In this section I will explicate my use of terms like 'environmental problems' and 'sustainability'. I will also present my own interpretation of the relationship between the free market and sustainability. In doing so I am forced to make choices, each of which could be the subject of a research project in its own right. I will keep the arguments in support of my choices to a minimum. My main concern is simply to make my choices clear, not to give a detailed justification for my approach. In the context of this study that would be going too far. Besides, my choices are fairly standard ones.

Environmental problems

As an ecological phenomenon, I define environmental problems as the loss and extinction of non-human life, as well as the loss and disappearance of the ecosystems of which this life is a part and on which it depends for its survival. In addition to this, environmental problems also take in the loss of quality of the natural environment of present and future generations of people, as well as the resulting decline in people's health and quality of life. In adopting this description of environmental problems I am following in the footsteps of *Caring for the Earth. A Strategy for Sustainable Living*, the report by the International Union for the Conservation of Nature and Natural Resources (1991). Like this report, I wish to emphasize that environmental problems do not only concern the survival of future human generations. Environmental problems are also concerned with non-human life and with quality of life.

One difference between my approach and that taken in *Caring for the Earth* is that I do not give a prominent place to the exhaustion of non-renewable resources.

As things stand, I think that authors such as Simon (1981) and Simon and Kahn (1984) have the better part of the argument with regard to the issue of the depletion of non-renewable resources. Simon challenges the common argument used by environmentalists that the depletion of non-renewable resources such as oil, copper and tungsten will seriously impact on the economies of the world (and therefore life as we know it) within five decades or so. Simon states that the increasing scarcity of non-renewable resources will at worst lead to a temporary rise in the price of these goods. This, and the foresight of some market parties will then spur an innovative search for alternatives. Eventually this search will substantially reduce or perhaps even completely do away with the need for the non-renewable good. One might of course call this argument overly optimistic. Nevertheless recent history seems to provide enough justification for it. Copper and tungsten, for example, were quite scarce a few decades ago. Now they have been replaced by other materials in many applications. With regard to oil, it is interesting to note that a major corporation like Shell is currently investing heavily in renewable resources such as solar energy. Shell knows that one day there will be no more oil and it is preparing itself for that day.

As a sociological phenomenon the environment is often described as a collective problem. Collective problems cannot be reduced to the sum of individual problems and they transcend the individual's options for action. When faced with collective problems *'the individual by himself is helpless and useless'*, as Dewey (1905d: 463) puts it. Dealing with collective issues therefore calls for collective action. Environmental problems are clearly collective problems. Whether we are dealing with the greenhouse effect or acidification or eutrophication or water depletion, all these issues are bound up with interdependent social processes in terms of their cause and effect. A lone individual is powerless to counter such processes. The solution calls for a coordinated effort.

Sustainability

I define the solution to environmental problems as sustainable development. This concept can be given substance in terms of material attributes or in terms of processes. Materially a society achieves sustainable development when it is organized in such a way that health, quality of life and the conditions for the life of present and future human generations are protected; a society that develops sustainably is also able to preserve present and future non-human life and its habitat (i.e. the natural environment). What this entails for the modern economy in practical terms will differ per product and production process. Some products, like dangerous pesticides, will have to disappear completely. However, most products or processes will have to fulfil specific requirements. Take, for example, sectors like the fisheries or car manufacturers. A sustainable fisheries sector will have to make drastic changes to its fishing techniques and introduce meaningful fishing quotas. Sustainable cars will have to meet stringent energy-efficiency demands, be built in modules and keep undesirable emissions to an absolute minimum.

In this study I will focus primarily on the products and production processes which need to be changed or scaled down. This will allow us to define sustainable development in terms of processes. Sustainable development means taking the need for sustainability into account in every phase of a product's life cycle (see for example Mol and Spaargaren, 1991b or Cramer and Zegveld, 1990: 391-410). The market has to modernize ecologically.

> '...(S)uccessful long-term environmental policy emerges as something more than additive end-of-pipe treatment fostered by successive environmental institutions. What is needed is a 'revolution' in technological efficiency and a radical change of production structures ... Sustainable development requires the full internalization of responsibility for environmental damage in the productive sphere; (it requires) a process of "ecologization"... .'. (Jänicke, 1997: 71-83)

The market and environmental problems

What is the underlying cause of environmental problems? In the literature we often read that environmental problems are a direct result of our materialistic, possessive, short-sighted and acquisitive Western culture and that this is a culture that 'we' are not really prepared to change (see Achterberg and Zweers, 1986; Van de Wal, 2001: 98-122). Without wanting to wage war on the relevance of such cultural visions, I feel that this kind of approach is too quick to assume that processes in society are directly driven by 'our collective will'. Environmental problems are not the consequence of the sum total of our individual wills, but the result of the sum of our actions. This last sum is not the direct result of a collective will but the sum of individual wills that have to manifest themselves within a specific institutional context. This institutional context for action rewards certain choices above others and can even be said to mould the will to a certain extent. It does not form an objective framework for the manifestation of 'our' will. No fisherman looks forward to the extinction of a fish species which he depends upon for his livelihood. Yet fishermen the world over are currently contributing towards the ecological exhaustion of the world's seas.

Another oft-mentioned cause of environmental problems is the rising world population (Ehrlich and Ehrlich, 1972). Once again let me make it clear that I am not out to belittle this argument. Nevertheless, I do not think that the increase in the world population should be top of the list of causes. One argument for this position is that in wealthy nations, such as Japan and the Western democracies, environmental problems are much more closely related to the increasing burden that each individual places on the environment than to population growth in these countries (Opschoor, 1989: 28-29). A second argument is that the poorer nations, often referred to as Third World countries, have not only had to face population growth in the last 100 years but have also had to deal with serious attacks upon existing institutional structures. It can be argued that the serious environmental problems in these countries should primarily be associated with this last cause. Bromley (1991) gives telling examples to support this view. He shows for example that the depletion of common pastures in some parts of Africa is caused by the fact that the traditional institutional structures for controlling access to these pastures are not suited to certain new situations. One of these is the situation in which relatively rich labourers return to the village of their birth to claim their pasture rights after working in the city or in another country for many years. The traditional system that grants each villager certain rights cannot cope with the large herds these men can afford as a consequence of their years of labour outside the village.

Another attack on the existing institutional structures described by Bromley has to do with the process of state-building that took hold of Africa in the 20th century. The newly emerging states often did not tolerate traditional, local governance structures. These competitors were destroyed, but many states could not fill up the vacuum that this created. The local anarchy that erupted as a result caused the depletion of many pastures. The demise of traditional governance structures meant the disappearance of controls on access.

This study assumes that the free market is the most important driving force behind environmental problems. This is a choice, as the above alternatives show, but it is one with a broad support base. Many academic analysts relate environmental problems to the workings of the market mechanism in one way or another (Ophuls, 1977; Opschoor, 1989; Heilbroner, 1975 and 1993; Harmsen, 1974). Sometimes it goes by another name, such as the Science-Technology-Capitalism system (Vermeersch, 1990).

In the rest of this study, the relationship between market and environmental problems will be explored in greater detail. For now I will make do with a general outline of why today's free market acts as the driving force behind environmental problems. The free market is a place where people operate within a specific institutional context. Two important aspects of this context are that people are concerned with their own interests and have to operate in a competitive atmosphere. This means that the market is selective in the kinds of products that it normally produces. The market excels in the production of certain goods, especially the kind that can be consumed by the individual consumer. The market is much less suited to producing other kinds of goods. In some cases it may even be said to be entirely unsuitable. Collective goods belong to this category. I define collective goods as goods that by their nature require coordinated action in order to be produced adequately. These goods therefore cannot be adequately produced on the market. There, actors operate independently of one another, and so they should, in order to keep competition alive. It is important to note that this definition differs from the standard economic definition. I regard the standard definition as defective because it stresses the consumptive aspects of collective goods and is blind to goods that are collective because of the peculiarities of their production, in particular the aspect of 'jointness of production'.

Collective goods can be further categorized by examining the reasons why their spontaneous production is obstructed. This allows three main types of collective goods to be distinguished. (1) First of all there are goods which can only be produced by collective effort. I will refer to the problem associated with this as *'jointness of production'*. The production of 'silence' fits the description of the first type of collective good. (2) Then there are collective goods which are not produced because each person calculates for themselves that the most rational option is not to contribute towards the production of the good but to hope that others do so. In this way they can benefit from the result without having to invest. In the theory of collective action this situation is known as the *prisoner's dilemma*. A variation on this is the situation in which individuals are, in principle, prepared to pay for the production of a collective good but ultimately do not do so because they do not want to play into the hands of *free riders*. (3) In the last category of collective goods, production is impeded by the fact that actors who are willing to take part have insufficient guarantees that all other actors will also contribute towards the production of the good. As a result they themselves ultimately refuse to contribute. No one wants to be the beast of burden for the rest. In the literature this is known as the *assurance game* (Sen, 1967). This situation occurs, for example, when willing businessmen are not prepared to go to any expense for a cleaner environment because they are afraid that they will lose out in terms of competitiveness if others do not do likewise. Another standard example of this type is that of dyke-building. Villagers who construct a dyke to protect their houses and land against floods cannot exclude non-contributors who may also live in the village. If the number of free riders is too great, it can put the construction of the dyke in jeopardy.

Environmental goods are usually seen as collective goods. The reason for this can best be illustrated by way of examples. Protecting an animal species always requires jointness of production. Small groups of poachers can hopelessly frustrate large-scale attempts to protect the white rhinoceros from extinction. Meanwhile the solution to problems like the greenhouse effect and acid rain are dogged by the problems of free riders and mutual assurance.

1.3 LIBERAL DEMOCRACY AS CONTEXT

In this study we will compare two mental models of social organization in the light of the liberal-democratic tradition. It is therefore important to stress that liberal-democratic thinking as such is potentially compatible with various mental models of social organization. There is no one single most suitable mental model of social organization to be derived from liberal-democratic thought. First of all there are significant differences of opinion within liberal-democratic thought about how its values should be prioritized and interpreted. While one author maintains that we should interpret freedom as negative freedom, another latches on to the concept of positive freedom. One academic might regard freedom alone as being of real importance and interpret equality in terms of equality before the law (equity). Meanwhile another might attach equal weight to freedom and equality and interpret the latter as equality of opportunity or even as equality of resources.

Secondly, it is true to say that even more or less identical interpretations of values can still give rise to major differences at the level of the mental model of social organization. For example, someone who gives prominence to freedom may argue that the market should be a sphere of limited responsibility where actors may exercise their rights. Another person, however, might just as easily state that freedom implies responsibility and argue that the market should therefore be a sphere in which actors take a certain degree of public responsibility upon themselves. These views are not necessarily different in terms of how they perceive the relationship between freedom and responsibility: a supporter of the first interpretation might feel that it is better to organize exactly the same responsibility collectively and indirectly and therefore via the state.

An important aspect of my research will be to analyse the indirect and direct responsibility models with regard to their respective views of the relationship between market, state and civil society. By way of introduction I will describe briefly how the liberal-democratic tradition looks upon this relationship in general. It is important to bear in mind, however, that it is hard to give a neutral description of the general relation between market, state and civil society within this specific context. Any such description is likely to favour the indirect responsibility model, since this model is still the dominant one.

The liberal-democratic tradition interprets modern society as a differentiated order. Society can be divided into three separate spheres of action or fundamental institutions which can and ought to be distinguished from each other. These fundamental institutions are the market, the state and civil society. Each of these fundamental institutions has its own logic, its own organizing principles, its own way of coordinating action and so on. From a functional perspective, each sphere also has its own tasks and responsibilities (although the indirect responsibility model clings more to an exclusive distribution of tasks than the direct responsibility model). The market is the sphere of economic production. As Samuelson and Nordhaus state (1985: 24), it serves to solve the basic economic questions of what, how much, how and for whom to produce. In addition to this, it is the sphere where people can experience their freedom. Mainly due to its unintended consequences, this sphere has a large impact on publicly relevant processes but it is not responsible for them.

Within liberal-democratic thinking, the state can be divided into a judicial and a political-administrative section. The judicial section is responsible for ensuring a proper dispensation of justice and takes in all the bodies directly associated with this task. Our study is not concerned with this part of the state. Within the liberal-democratic tradition, the political-administrative section of the state deals with public issues. This part of the state can further be divided into a political and a bureaucratic section. The political section covers all bodies whose representatives are elected, either directly or indirectly. The bureaucratic part covers the entire spectrum of support organizations whose members are appointed on non-political grounds. It is unclear exactly what status should be accorded to political parties within the liberal-democratic tradition (Ankersmit, 1995). As I see it, they should be regarded as bodies from civil society operating within the state.

In their standard work, Cohen and Arato (1992: ix) provide us with a good working definition of civil society: '*Let us start with a working definition. We understand* 'civil society' *as a sphere of social interaction between economy and the state, composed above all of the intimate sphere (especially the family), the sphere of associations (especially voluntary associations), social movements, and forms of public communication*'. Within the indirect responsibility model, civil society occupies a position comparable to that of the market when it comes to political-administrative or public issues. It influences publicly relevant processes and may even create new problems, but it does not provide solutions to public problems. The direct responsibility model seeks ways to involve civil society in controlling the market.

The distinction between the public and the private

In the previous section I characterized environmental problems as collective problems. It is necessary to interpret this conceptualization from a liberal-democratic perspective. Liberal-democratic thinking is essentially normative thinking about individual freedom and/or self-determination. Therefore, the most basic categorization of problems is not individual versus collective, but private versus public; that is to say the classification of issues as those that people may and should handle by themselves or those that they may and should handle as citizens. However, the sociological dichotomy between individual and collective does not coincide with the divide between the private and the public. For the sake of conceptual clarity, we have to determine the relationship between these two dichotomies. The context of this research thereby justifies paying some extra attention to the intricate dichotomy of 'private versus public'.

Within liberal-democratic thinking the distinction between public and private is a distinction with several layers of meaning. I distinguish between public and private as denotations of specific *domains of action* and public and private as denotations of specific kinds of *issues*. As domains, the distinction between the public and the private refers to spheres of action in which the state has specific rights, duties, powers, liberties and immunities in relation to individual citizens, and vice versa. The *public domain* is the domain in which the state is entitled to act (although the manner in which the state is permitted to act can still be subject to conditions). The private domain is the domain in which an individual is free from intervention by the state or by other individuals. In the private domain, others have no say. Within modern Western societies, questions concerning the choice of one's friends and partner for example are considered private by all liberal-democratic theorists.

The 'public domain' and 'private domain' should not be interpreted as *spatial* areas in the literal sense. Such an interpretation inevitably leads to absurd conclusions. Private property is, for example, considered part of the private domain. If one interprets the private domain as something spatial then there would hardly be any public domain in contemporary society. Children, for example, would be at the mercy of their parents as long as they were actually present on their private property. It is therefore more reasonable to interpret the private and public domains as specific spheres of *choice*.

As issues, the distinction between private and public creates order in democratic society as a political arena. It can for example be used to delimit the responsibilities of government (though not its powers). Private issues are issues which only concern the individual. Public issues are issues that concern all. The implications of an issue being public are threefold. Firstly, the public nature of an issue means that actors have to view it to some extent from within their role as citizens. Moreover it implies that every citizen has to acknowledge that they have a responsibility to contribute to the solution of that issue. Lastly it implies that under some specific conditions the state has a legitimate right to regulate society in relation to that issue and coerce citizens into complying with the rules that it prescribes.

But when does an issue concern all? An issue can become public because the citizens or their representatives decide within the context of democratic bodies that it should be so. Issues can also become public because they should concern everyone by their nature. A number of comments need to be made to accompany this definition of public issues. To begin with, let us focus on the notion that an issue can become public because the citizens decide it should be so. Within liberal-democratic thinking, the right of the citizens to declare an issue public is strictly limited. Individual citizens also have rights and, in certain instances, some of these rights are more fundamental than the collective right to declare an issue public. Some issues are therefore necessarily private, for example, the issue of whether a specific individual should become a butcher or a baker. Furthermore, the right to declare an issue public is sometimes also conditioned by fundamental obligations that are imposed on individuals within liberal-democratic thinking. One example is the duty all individuals have to take care of themselves (barring exceptional circumstances). This also places limits on the constitution of public issues.

Secondly, it may sound a little strange to regard some issues as public because 'they *should* concern everyone'. This formula takes its rationale from the notion that liberal democracy defends specific values. These values need to be maintained in order to guarantee the future of liberal democracy. Accordingly there are some issues with regard to which the citizens do not have the right to determine public status. Normatively speaking such issues achieve public status by virtue of the logic of liberal-democratic thinking itself, as the following example illustrates. The legendary civilization of the Ancient Greeks mainly left moral issues concerning slavery and infanticide to the free choice of the head of the household. The liberal-democratic tradition abhors both practices. All right-minded liberal democrats therefore not only seek the freedom to disassociate themselves from such practices but also refuse to be part of a society in which such practices take place. In short, the logic of liberal-democratic thinking makes slavery and infanticide public issues.

The relationship between the public domain and a public issue

How are the different layers of the concepts 'public' and 'private' related? There is no general answer to this question. This is precisely the point at which the indirect responsibility model and the direct responsibility model diverge. A typical feature of the indirect responsibility model is that its supporters superimpose the two components of the dichotomy on one another. They bring together 'public and private as denotations of specific domains of action' and 'public and private as qualifications for specific kinds of issues'. This has several implications:

* Within the indirect responsibility model the private domain is conceptualized as the domain in which actors do not have responsibility for public issues. As long

as one does not superimpose the two components of the private and the public, this conceptualization is not necessarily logical or inevitable. It is conceivable that a person within the private domain has some responsibility for public issues. The only condition is that we drop the idea that the state or other citizens are entitled to pressure a person into acknowledging this responsibility and acting accordingly.
* The indirect responsibility model has a tendency to reduce the responsibility of individuals (in the private domain) to the responsibilities that can be exacted from them. This implies that the responsibilities of a person tend to coincide with the duty to obey the law and the rules of common decency.
* The indirect responsibility model has a tendency to interpret the public domain exclusively as the domain of the state.
* The indirect responsibility model has a tendency to conceptualize public issues as issues that concern the state exclusively.
* The indirect responsibility model has a tendency to look upon the state as the only or at least the prime public actor.
* Within the indirect responsibility model, the proposition that an issue is 'public' warrants both the conclusion that the state has a responsibility and a legitimate reason to regulate society.

It is characteristic of the direct responsibility model that the two components of the dichotomy are not seen as fully overlapping. Common propositions within thinking on corporate social responsibility bear this out. One example is the proposition that actors on the market - and thus acting within the private domain - have a moral responsibility to take action to combat the pollution they cause, even if they are not legally obliged to do so. The separation of the two components of the dichotomy also carries a number of implications:
* Within the direct responsibility model the private domain is regularly conceptualized as a sphere where actors do have some responsibility for public issues. A typical feature of this public responsibility within the private domain is that people are morally required to acknowledge it but that the state or society at large cannot coerce them into doing so.
* The direct responsibility model has no tendency to reduce a person's responsibilities to those that the state can enforce.
* The direct responsibility model has no tendency to conceptualize public issues as issues that concern the state exclusively.
* The direct responsibility model has no tendency to look upon the state as the only public actor, although it does not necessarily deny the crucial importance of the state for dealing with public issues within a modern context.
* Within the direct responsibility model the proposition that an issue is 'public' only warrants the conclusion that the state probably has a responsibility in relation to that issue. Whether the state has a legitimate reason to regulate society stands in need of a distinct legitimization.

'Public vs private' and 'collective vs individual'

This lengthy discussion on the subject of 'public versus private issues' arises from the need to determine the relationship between 'the private and the public' and the sociological distinction between 'individual and collective problems'. We can now conclude that theorizing in terms of the distinction between individual and collective problems is based on tacit knowledge which is not neutral in the context of this research. The distinction standardizes acting in ones self-interest (within the confines of the law) as rational for example. This implies that theorizing in terms of 'individual vs collective problems' tacitly conceptualizes the private domain as the domain of private issues. In other words, theorizing in terms of the distinction between individual and collective problems fits in best with the indirect responsibility model. From the perspective of the direct responsibility model, many connotations that are commonly linked to the idea of a 'collective problem' do not follow automatically. One example is the often-heard suggestion that collective problems belong to the public domain or that they legitimize state action. This bias inherent in the distinction between individual and collective problems where the issues addressed in this study are concerned implies that this twin concept should be employed as little as possible. I will therefore mainly conceptualize environmental problems as public issues. I cannot completely avoid speaking in terms of 'individual vs collective problems', however, as this would undermine my efforts to offer an accurate account of the thinking on environmental problems in three academic fields. Within these disciplines it is quite common to conceptualize environmental problems as collective problems.

1.4 LOOKING BACK AND LOOKING AHEAD

This study will examine two mental models of social organization. A mental model of social organization is a model which indicates what the relationship between the state, the free market and civil society should be like and which also addresses the main structural aspects of each of these fundamental institutions. This study looks at these mental models in the light of the question of how best to control the free market in present-day Western societies.

I have used the term *indirect responsibility model* to refer to the mental model of social organization that was most widely accepted in the 20th century. The main features of this model are the notion that actors in the market only have a limited responsibility for public issues, that the state has the task of solving public problems and that the state is only permitted to control the market by means of limiting conditions.

I refer to the alternative mental model as the *direct responsibility model*. This model has been worked out in less detail than the indirect responsibility model, but has attracted a great deal of interest in recent years, especially from administrative theorists and business ethicists. This mental model centres on the idea that actors on the market do have a certain responsibility for public issues and that the state is not the only institution capable of bearing public responsibility.

The advocates of the alternative model feel that the indirect responsibility model no longer forms an adequate frame of reference for thinking about control over the market. One of the reasons given in support of this position is that the state has become irrevocably overloaded within the indirect responsibility model. The criticism levelled at the indirect responsibility model is not always convincing, however. What is more, the critics of the indirect responsibility model often lose sight of the normative dimension. They forget that an adequate present-day mental model has to fit within the liberal-democratic tradition and that this tradition places considerable normative demands on the way in which society should be organized. These concerns give rise to the first key research question of this study: can it actually be demonstrated that the indirect responsibility model is no longer feasible? Is there enough support for the argument that the indirect responsibility model is outdated?

At present the direct responsibility model has not been extensively enough developed to be taken seriously as an alternative mental model. In addition to this, advocates of the direct responsibility model do not take full enough account of the liberal-democratic context. For example, they fail to pay sufficient heed to the liberal-democratic aversion to corporatism. The second key question of this study follows on from these observations: what form does a mental model take which has at its core the idea that market actors have to bear public responsibility? In other words, what does an adequate direct responsibility model look like?

Mental models of social organization largely exist as background knowledge. This fact makes it difficult to study mental models directly. Bearing this in mind, the initial focus of this study addresses a more general question: what approach do present-day Western societies take towards controlling the market?

The focus of this question has been narrowed in a number of ways. First of all, the question will be applied to a specific issue; sustainability. Next, the investigation only deals with how this issue is perceived within the academic disciplines of economics, administration and political philosophy. The analysis will further restrict its focus to a main stream within each of these disciplines. In economics, this will be the neoclassical tradition; in administration, thinking in terms of 'limits of state action'; and within political philosophy, the emergent tradition which places civil society at its heart.

In the three chapters that follow, I will outline these schools of thought within economy, administration and political philosophy. At the end of each chapter I will bring my findings back into direct relation with the key research questions. The final chapter is devoted entirely to the further development of the direct responsibility model.

CHAPTER 2

ECONOMIC THEORY:

The market as problem and solution

'The price mechanism - the instrument of the invisible hand - does, it is admitted, have some serious defects. But the remedy is not to abandon that mechanism or to superimpose other instruments with which it is not readily cross-bred. Rather, the thing to do is to use prices themselves, as far as possible, as the most promising means to cure their own shortcomings.

W.J. Baumol and S.A. Batey Blackman (1991: 47)

2.1 MARKET FORCES AND YET MORE MARKET FORCES

Concepts such as 'market forces' or 'pricing' occupy a prominent place in present-day thinking about sustainable development. From all quarters we are assailed by the message that we should be making the greatest possible use of the market in order to realize this public objective. In this respect, the spirit of the age is sure to bring a warm glow to the cheeks of many a neoclassical environmental economist. For decades environmental economists have maintained that the state manages society in a way that is far from judicious. It employs sanctions and regulations to deter organizations from pursuing activities that are potentially profitable for them and/or forces them to undertake activities which can only work to their disadvantage in economic terms. Economists argue that it would be more astute and even theoretically compelling to adopt a less combative system (Marcus, 1982: 173). The context in which businesses operate should be shaped in such a way that the distance between profitable actions and actions which benefit the public interest is reduced as much as possible. The price mechanism must be harnessed for the good of government. This would allow the economy to be managed both more effectively and more efficiently.

In the late 1960s and early 1970s, when the first wave of environmental awareness and environmental legislation swept across the globe, the economists' view of these matters failed to make much of an impression on the administrators, the environmental lobby or the public at large. Thomas Schelling (1983: ix-x) gives a highly plausible account of the frustration or indeed the astonishment experienced by the economists when he writes:

'There is a discrepancy between the approach of economists to environmental protection and the approach of nearly everyone else. ... Though economists

23

acknowledge that many environmental impacts cannot best be managed through the price system, pricing is their first choice among management techniques. ...In politics, economists propose but noneconomists decide, and prices - or "incentives" generally - are not characteristic of governmental intervention. ...

There are two polar possibilities here. One is that economists exaggerate grossly the virtues of the price system in environmental protection, underestimate egregiously the difficulties of implementation, and are bemused by their own theoretical constructs. The other is that they have failed to get their message across, or their audience is perversely or irrationally predisposed against their ideas, or there is some other removable impediment to the initiation of wise policies'.

Present-day economists can no longer complain of a lack of attention or a dearth of support. The market now takes pride of place in the debate on managing society. Even in the more specific discussion on sustainability, the neoclassical environmental economists can count on receiving serious attention. Pricing, for example, is an important element in the thinking behind Dutch environmental policy (Tweede Kamer, 1989; Ministerie Volksgezondheid, 1993 and 1998).

Four interpretations of 'market-based regulation'

If we were to boil the neoclassical environmental economists' strategy down to a single statement, we could describe it as an attempt to beat the market at its own game. The market has to be made more sustainable by harnessing the regulatory effect of market forces. Given the central place that this concept of 'market-based regulation' occupies in this chapter, I would first like to discuss the term itself in greater depth.

'Market-based regulation' is not a term that is very widely used. I employ it here to denote the action of resorting to the market as a regulatory mechanism, as expressed in ideas like 'pricing'. However, it is important to realize that both within and beyond the economic tradition, an appeal to the market as a mechanism for coordinating public matters is far from being a clear-cut undertaking. From an analytical point of view, there are at least four different definitions of the concept of market-based regulation. Making a clear distinction between these various meanings in practice is often far from easy.

The first explanation of the concept is the most moderate when seen from an administrative point of view. It requires that knowledge of how market processes work and the way people operate in the market be taken into account as much as possible by public administrators when they develop policy. Wherever possible, the state should make use of market forces. The state should encourage market agents to act in an environmentally responsible way by means of economic incentives (money). Businesses should come to view the environment as an economic variable through which the company can cut costs or accrue benefits.

I refer to this interpretation of market-based regulation as 'management according to market principles'. Management according to market principles does not necessary result in direct advantages for the government compared to a traditional licensing system. According to the environmental economists the main overall benefit is to the business community and society as a whole. While the traditional licensing system creates a situation in which every company has a fixed set of standards imposed upon it, the market principles approach gives each company the option of seeking out the healthiest balance between investing in environmental measures and paying the price of polluting the environment. Another advantage of the market principles approach is that it encourages companies to set out on the path of ongoing ecological modernization. Since this approach makes an economic variable of the environment, businesses will always have an incentive for exploring possible advantages to be gained in that area. This is something the application of traditional, fixed standards cannot achieve. Under such a regime, companies have no economic motivation for doing more than the minimum needed to obtain their licence.

Typically, this interpretation of market-based regulation is not presented as a self-contained alternative to traditional ways to govern: the strategy only constitutes an *addition* to the existing range of measures. The tasks and responsibilities of the state have not been tampered with. The state is only called upon to deploy the means at its disposal with greater perspicacity. To my mind, the current arguments in favour of 'tradable emission rights' are a form of management according to market principles (Van Duijse and Nentjes, 1998). When all is said and done, a tradable emission right amounts to no more or less than a special type of licence, with the special feature of tradability ensuring an extra measure of efficiency.

In its second sense, the drive towards market-based regulation is presented as a fully fledged alternative to the idea that all public objectives and all public policy should be realized by the state. Advocates of this interpretation bemoan the overloading of the state and see this as partly due to the problems inherent in a bureaucratic or hierarchical management style. In order to improve the way society is governed, tasks and public responsibilities need to be transferred to the market as much as possible. The market should become a place where public objectives are met.

Advocates of this second explanation feel it is high time that the deeply ingrained association between the state and public responsibility was severed once and for all. The market must become the mechanism through which public goals such as sustainability are realized. As C.L. Schultze (1977) so neatly states in the title of his book, it is all about making 'public use of private interest'. Another way of putting it would be to say that this approach to harnessing market forces involves internalizing tasks that were initially external to the market process (Dragon and O'Connor, 1993:127), thus enabling them to be fulfilled by the market. Imposing duties on fuel and levies on the use of groundwater are two examples of such an approach. In the case of the former, a price is put on pollution while in the latter, people are required to pay for using a scarce environmental resource. Since today's market has failed to meet these public objectives, it is clear that this form of market-based regulation necessarily involves drastic changes to the internal structure of the market and/or far-reaching expansion of its domain.

The third approach to harnessing market forces is the most radical in normative terms. Here too, market-based regulation is presented as a fully fledged alternative to state control and once again we encounter the notion that the state is overloaded and that there should be a transfer of responsibilities between the state and the market. However, the supporters of this approach do not advocate changes in the structure of the market. What is more, they see the transfer of responsibilities not simply as a means but as an end in itself. Preference is given to market-based regulation on the largest possible scale, even if this means that certain public objectives (e.g. sustainability) are not realized. From this point on, I will characterize this view of market-based regulation as the drive towards a neoliberal market (Nozick, 1974; Hayek, 1976; see also De Beus, 1989: 359-430; Cohen, Rogers and Wright, 1995: 14-21).

The neoliberals have a less instrumental view of the market than advocates of the second view of market-based regulation. From a neoliberal perspective, the market is not simply a handy mechanism which can and should be judged in terms of the targets it reaches. For neoliberals, the order of the market is an inherently good order (Hayek, 1976; see also Kymlicka, 1995: 95). Accordingly, neoliberals refuse to set out specific goals for the market to attain. As long as organizational principles like that of competition are in place, any result of the market process is acceptable and correct.

In the fourth interpretation, the concept 'market-based regulation' means that *actors* operating in the market should directly and consciously take some responsibility for the public effects of their actions. This interpretation of market-based regulation demands that market actors operate in the market in such a way that they take political, administrative, normative and/or social considerations into account when reaching their decisions. On the grounds of their responsibility, it can and should be possible to demand that actors do things that may go against their own rational, economic interests.

The way in which neoclassical environmental economists (Freeman III et al., 1973; Baumol, 1975, 1991; Herfindahl and Kneese, 1974; Kneese, 1977; Schultze, 1977; Tisdell, 1993; Pearce et al., 1990; Van Ierland, 1990; Neher, 1990; Pearce and Turner, 1990; Pearse, 1994; See also: Arrow, 1974 and Schelling, 1984.) see the concept is a mix of the first and the second interpretations detailed above. On the one hand, market-based regulation is seen as a *handy instrument* which the state could use in its efforts to manage the market. On the other hand, it is regarded as *an alternative to* state control. Neoclassical environmental economists feel that the drive towards sustainability has to be anchored in the structures of the market in such a way that economic actors will take it into account as a direct result of their pursuit of profit. The second account of the concept is often the more dominant (see also Section 2.4). From a theoretical point of view it is also the more interesting: it opens the door to a society in which the state can transfer a share of its tasks to the market. With this in mind, I will focus on this second notion of market-based regulation for the rest of this chapter.

The second and the fourth interpretation of market-based regulation sometimes get mixed up, as is the case in Dutch environmental policy (Ministerie Volkshuisvesting, 1998). For the understanding of this study it is important not to make the same mistake here. The essence of market-based regulation in the second (neoclassical) sense is the notion that it is both possible and desirable for the market to generate a structure of economic stimuli to provide actors with the *economic* motivation to undertake activities which are considered desirable from a public perspective. Actors aim for sustainability, not for its own sake but for the sake of their own economic interests. In the words of the famous phrase: 'private vice, public benefit'. A plea for market-based regulation in its fourth guise is actually a plea to partially disengage the rationality of the market. To paraphrase the celebrated saying above, it amounts to 'private responsibility, public benefit'. We might also say that the second interpretation of market-based regulation is consistent with the indirect responsibility model, whereas the fourth interpretation captures the essence of the direct responsibility model.

Structure of the chapter

In this chapter I wish to analyse the thinking of today's neoclassical environmental economists on controlling the free market. In Section 2.4 I describe the specific manner in which neoclassical environmental economists employ the concept of market-based regulation in their political-administrative diagnosis of modern-day society and their suggested treatment of its ills. Sections 2.5 and 2.6 contain an analysis and evaluation of this train of thought. However, before exploring these areas I would first like to focus on the concept of the market itself. In Section 2.2 offers a brief sketch of the way in which academics outside of the neoclassical tradition think about the market. In Section 2.3 I will go on to describe the neoclassical concept of the market. Returning to the present section, I would like to conclude by making a few short remarks and discussing a number of methodological stumbling blocks.

My first remark concerns the relationship between neoclassical environmental theory and the neoclassical tradition in general. The thinking of the environmental neoclassicists is strongly influenced by the neoclassical tradition as a whole. In order to obtain an accurate impression of the mental model of social organization adopted by the environmental neoclassicists, I am therefore obliged to make numerous forays into the neoclassical tradition in general.

My second remarks touches on economic theory outside the neoclassical tradition. I do not wish to ignore this category of economic theory entirely. A brief account with reference to a number of authors outside the neoclassical tradition can help to foster a greater insight into the market and will also prove useful when it comes to putting neoclassical thinking in its proper perspective. I will place all of these theories under the collective heading 'non-mainstream economics'. However, I am very much aware that this does not do justice to the variety of economic theory outside the neoclassical tradition.

Methodological stumbling blocks

Methodological discussions, as they take place within the academic discipline of economics, should not really form part of this study. Nonetheless it is important to identify a number of these discussions here. The first concerns the relationship between model and reality in the neoclassical approach. The model-based approach is one of the most characteristic aspects of the neoclassical tradition.

> 'The neoclassical school is a broad church, offering a methodology and paradigm embracing many sects. The high priests of the church are well versed in mathematical technique, which they employ to trace out the consequences of individual behaviour on the assumption that economic agents constantly strive to maximize their economic well-being. These agents may not be, indeed typically are not, regarded as flesh and blood actors; they are mythical creations, designed so that their behaviour is perfectly predictable according to hypothetico-deductive chains of reasoning.
>   On the other hand, neoclassical economists would claim that: first, though the behaviour patterns assumed of these invented individuals, the invented economic agents, may not at all reflect the rich complexity of inconsistencies and uncertainties of human behaviour, yet the maximizing assumption recognizes one very important component of such behaviour, so that the conclusions reached are of practical relevance; second, only a rigorous methodological approach, as exemplified by mathematical techniques, can assure that the conclusions reached are not logically erroneous.' (Alec Gee, 1991:71).

Theory formulated on the basis of models is potentially very powerful, especially when academics feel the urge to make predictions. However, there are two methodological questions that an academic who uses models must always keep in mind. The first question is: how do I get into my model (from reality)? The second question is: how do I re-emerge from my model (and back into reality)? In short, model and reality have to be clearly distinguished from each other at all times. The ongoing objection raised by non-mainstream economists and others academics is that, in the course of the 20th century, the neoclassical tradition has lost sight of the distinction between model and reality: the two have merged inextricably (Breiner, 1995; Hayek, 1948; Knight, 1935). A typical symptom of this problem is that the market is no longer seen as a construction, the product of institutions like the law (Commons, 1922), but is instead presented as a natural phenomenon.

As a consequence of this methodological laxity in the neoclassical tradition, the outsider who wants to give fair coverage to neoclassical theory while paying due deference to fundamental methodological principles sometimes finds himself torn between the two. While I have made a genuine attempt to span this divide, I fear it has not always been possible to pull it off and that neoclassical economists may sometimes feel the urge to contest my interpretation of their vision.

The second discussion centres around the question of whether the academic discipline of economics is a normative discipline. Is economics based on normative principles? Non-mainstream economists and economic philosophers tend to answer this question in the affirmative (Knight, 1935; Sen, 1987; Hausman and McPherson, 1993). Frank Knight (1935: 19) for example states that *'Economics and ethics naturally come into rather intimate relations with each other since both recognizedly deal with the problem of value'*. Neoclassical economists meanwhile have been resolutely shaking their heads at such statements for almost a century, with a few notable exceptions such as Marshall.

Pigou sums up the neoclassical viewpoint neatly: *'(Economics) is a positive science of what is and tends to be, not a normative science of what ought to be.'* (Pigou, 1920: 5). In my opinion, this neat positivist division between science and ethics cannot be maintained, especially not when it comes to neoclassical theory. As the previous chapter revealed, neoclassicists like Pigou view the market as a functional sphere. This in itself is a significantly normative point of departure. The indebtedness to norms is also brought into sharp relief by asking a question like why a great many economists regard the Pareto optimum, for example, as desirable.

In my opinion, the fact that neoclassical economists continue to resist the idea of a normatively charged discipline, despite a century of convincing evidence from Knight to Sen, can only be understood as a consequence of the dominance of positivism in the 20th century. This philosophy of science has ruthlessly separated norms and facts, ethics and science. Faced with the supremacy of this philosophy every self-respecting academic must object to any suggestion that his discipline could possibly be normatively tainted. His academic integrity depends upon it.

Be this as it may, the steadfast refusal of neoclassical economists to acknowledge that their discipline has anything at all to do with norms clouds communication with other academics. The latter are called upon to defend their position again and again. As a result, we often encounter sentences like the following:

> *'Many economists who regard themselves as neoclassicists would deny that their analyses ... carry any normative implications, meaning a value judgement of what* ought *rather than what* is. *... Such a position cannot be entirely well founded'* (Alec Gee, 1991: 74).

On occasion I too will develop an argument which presupposes that the academic study of economics is to some extent a normative discipline. However, it would be cumbersome to have to interrupt the flow of my argument at every turn in order to launch into a 'metadiscussion' on this issue. I will therefore assume that, by means of this brief methodological intermezzo, I have laid my cards firmly on the table.

The third discussion pertains to the commensurability of all things of value. Two goods are said to be commensurable when the value of one good can be more or less completely expressed in terms of the other. It implies that a common scale is available on which the value of the two goods can be adequately expressed. Incommensurability means that this is not possible. By way of example, a computer and a steak are commonly regarded as commensurable, at least as long as they are both for sale in a shop. The value of both goods can be entirely expressed in terms of money. My son, however, is not commensurable with a computer, since there is no amount of money for which I would wish to sell him. In more general terms we can say that, since the abolition of slavery, modern Western liberal society regards human beings as incommensurable with other values or goods (or even with each other).

It is important to realize that commensurability is not the same thing as comparability (Bernstein, 1983). If my house catches fire, the dilemma of whether to save my son or the computer is hardly likely to give me much pause for thought. Although people cannot be bought and sold, there are certain conditions (e.g. when dangerous work needs to be carried out) under which a decision still has to be taken as to what constitutes an acceptable level of safety. This means that in practical terms it is both possible and necessary to compare incommensurable goods on a rational basis. However, such comparisons are often difficult to make precisely because there is no clear scale available upon which such a decision can be based. Deciding between two incommensurable factors involves comparing values of a very different nature (economic, political, normative, aesthetic and so on) and weighing their merits against each other.

The neoclassical tradition - environmental economists included - takes a notorious standpoint regarding the incommensurability of values and goods. I employ the term 'economism' to describe this position. My use of this term is based on two assertions. On the one hand economism is the conviction that goods and/or values can only be compared against a clear and unambiguous scale of measurement. On the other hand, economism maintains that every objective, every value and/or every good can be transformed into an economic good, and that for this reason all human objectives, values and goods are commensurable and comparable in relation to the measuring rod of money (or other single-scale, quantitative, market-related measuring rods, like quantification in terms of utility units).

Neoclassical economists therefore deny the existence of incommensurable goods. In their eyes, all value attached to things can be expressed on an economic scale. But economism is not the same as neglecting or ignoring values and goods which people normally view as 'not economic', such as the beauty of a landscape and the like. To ignore non-economic goods is to give them no economic value. Economists on the other hand are convinced that scientific methods and procedures can be developed which will allow us to place all goods, values and objectives on a market-related or economic scale.

Economism has far-reaching consequences for neoclassical theory on controlling the market. It is a cornerstone of the argument put forward by many neoclassicists that under perfect conditions the market not only produces an economic optimum (see Section 2.3) but also a *social* optimum. For the sake of argument, let's suppose for a moment that such a perfect free market can exist in practice. For the same reason, let's assume that all value is linked to *things* that can be *produced*. Based on these assumptions, the neoclassical conclusion that it is desirable to shape society in such a way that it resembles the free market as closely as possible, seems warranted. If all value is linked to goods, if all goods are economic goods, and if the market is the most appropriate mechanism for the production of economic goods, then the market is the best mechanism to coordinate *all* of society's goods. However, if one rejects these assumptions, there is no logical link between the argument that the market leads to an economic optimum and the position that all processes should be regulated by the market as much as possible. In that case the economic optimum and the social optimum become disconnected. From an academic point of view one would therefore expect the neoclassical economists who embrace this view to go out of their way to produce a meticulous argument convincing us of these assumptions. This expectation is disappointed, however. Neoclassical economists hardly ever defend their starting points (at least not without determining their conclusions in advance). I think their economism hampers them in this respect. Economism is never a conclusion of their argument but always an assumption.

Neoclassical environmental economists usually have an economistic mindset. As a result, their leaps in argumentation sometimes appear odd to a non-economist. The ease with which environmental economists exchange 'the economic optimum' for 'the social optimum' is one such instance. Having discussed economism at length here, I feel justified in leaving these strange leaps for what they are during the rest of my analysis. Like the neoclassical environmental economists, I will skirt around the economism discussion. My reason for doing so is that, in this chapter, I am interested in the neoclassical mental model of social organization.

It is probably clear from the above that I do not share the economistic mindset of the neoclassical tradition. So as not to leave this criticism hanging in mid air, I will briefly present a number of major objections to economism, all of which have been around for some time. Economism has been heavily criticized throughout the 20th century by academics of all kinds including many non-mainstream economists (Hirschmann, 1992; Walzer, 1983; Knight, 1935; Etzioni, 1988; Robinson, 1962). Knight (1935) for example states that economism reduces human beings to mere consumers, thereby ridiculing the truly valuable aspects of life. The value of being human does not lie in consumption but in discovering value and striving towards its realization. In short, economism is based on a fundamentally flawed vision of humanity. Another important point raised by Knight is that every economistic analysis must proceed from the idea that human needs have the status of fixed scientific data. However, this is simply not the case. One of the main characteristics of human nature is that needs change. Human beings are creatures who not only think about means but who also reflect on ends.

Robinson (1962) describes economism as a self-defeating strategy. The aim of setting out all values on an economic scale can only make sense if we know what the economic scale actually means. Yet the neoclassicists do not have an adequate theory for translating 'economic value' back in terms of 'value'. Of all the theories which exist in this area (labour theory of value etc.), the most consistent and accurate is the one which says that no unequivocal relationship can be established between economic concepts like 'price' or 'utility' and the concept of 'value'. A price only has meaning within the system of prices but not outside it. Some objects of value are of limited utility in economic terms or are low in price. Other not so valuable products have an enormous economic value attached to them. In short, the lack of an adequate theory to link economic value to value in general renders meaningless the economistic attempt to express all that is valuable in economic terms.

The work of other authors, like Sen (1990; see also Etzioni, 1988), brings us to another serious criticism. Sen argues that a person is driven by incommensurable *kinds* of motivation and goes on to distinguish preference, duty and loyalty as kinds of motivation. He claims that human behaviour cannot be adequately understood if one does not acknowledge the fundamental differences between these kinds of motivation. One of his examples is the conduct of a soldier who staunchly stands his ground under fire. It is very hard to explain such behaviour if the only category of motivations that one can draw from is 'preferences'. Sen's position implies that any quest in search of the one true scale of value is doomed.

An economist might object to this stance by reminding us that in practice we are forced to compare the incommensurable. So one way or another we have to come up with a single scale of measurement. With the help of Bernstein (1983) we can reply that 'incommensurability' and 'incomparability' are different concepts. With the help of Scanlon (1998) we can add that the judgement of reasons often implies more than just a one-dimensional measurement of their 'weight'. Some reasons have another function in our thinking. They structure our judgements, for example by conditioning the kind of reasons which are allowed (i.e. which constitute good reasons) in a specific context. One might imagine a situation involving a person who gets their kicks from beating stray dogs. In an economistic discourse it is hard to morally object to this conduct, unless we attach value to stray dogs. This becomes problematic in certain contexts, however, like the example of a neighbourhood plagued by packs of stray dogs. Scanlon's position helps in evading all paradoxes. Within his framework we can hold that people are morally obliged to not harm stray dogs without due cause. This way of thinking places certain pleasures of certain people out of bounds.

To sum up, we could say that, by conceptualizing many non-economic values in economic terms, the economistic perspective fails to recognize the unique character and nature of these values. The economic perspective is after all a perspective which actors adopt under specific circumstances, with specific motivation, with a specific disposition and with specific capacities (see Section 2.2). All cultural, normative, aesthetic, political and other values, which can only be experienced or fully experienced by adopting another role, by definition remain unexpressed or inadequately expressed within an economic perspective.

2.2 MARKET DESCRIPTION I: THE NON-MAINSTREAM APPROACH

The economist Geoffrey Hodgson (1988: 172) remarks, more or less to his own surprise, that '*definitions of the market in the economic literature are not easy to find*'. 'The market' as a concept is perplexingly ill-defined. A little later, Hodgson goes on to qualify his position: definitions of the market are not so much untraceable as unclear, vague and all too clearly inadequate (see also Coase, 1988: 1-7).

This study requires an adequate description of the free market. This description need not be overly concerned with the question of what this market actually *is* as long as it describes the ways in which the market is conceptualized within economic theory. This shifting of perspective simplifies the issue a little, but not by much. The same is true of the fact that we can limit our view to the modern free market within the context of a liberal-democratic society, such as the Netherlands, Britain, the USA or Germany.

Contemporary economic theory describes the modern market in many different ways. I have instilled a measure of analytical order into these myriad definitions by distinguishing between three types of market description: the *empirical-historical* description of the market, the *model-based* description and the *functional* description. The empirical-historical description is allied to non-mainstream theory, while the model-based and the functional descriptions can be found within the neoclassical tradition. These three market concepts will be discussed in the next two sections. I will conclude Section 2.3 with a brief discussion of the remarkably great tension that exists between these various conceptualizations.

An empirical-historical description

Outside of the neoclassical tradition, the market is often described along empirical-historical lines. The market is defined by summarizing its empirical characteristics (see for example Lindblom, 1976; Hayek, 1976; Koslowski, 1986; De Beus, 1989). Various authors list various characteristics. To my mind, these inventories can be distilled into a definitive list of seven characteristics presented below. Since not all of these characteristics are of equal relevance to this study, the coverage devoted to each will vary.

I. A decentralized order

Within non-mainstream economic theory the free market is first and foremost conceptualized as a decentralized order. There is no planner or authority to coordinate economic behaviour. Nor is there a legal power or institutionalized force to press actors into making certain choices. Action is coordinated on the basis of exchange between free individuals who are equal before the law (Lindblom, 1976).

In our everyday conversation, we are apt to associate 'exchange' with an activity that people carry out 'in complete freedom'. This association is just as readily made when it comes to thinking about exchange on the market, all the more so given the broad currency of the term *'free* market'. Within the context of our present discussion, this association is not appropriate. The fact that the market is not controlled by a concrete authority which forces people into making specific choices, is not the same as saying that there are no forces at work. On the market, people are coerced into action by economic forces. Whoever sets about resisting such forces may not run the risk of imprisonment but he will pay dearly for his choice, with poverty or bankruptcy as the ultimate punitive measures.

In this respect, it is important that we make a clear distinction between two types of compulsion. On the one hand there are forces which arise as a result of the power exerted by other actors in the market. Monopolization is a classic example of this brand of market power. On the other hand, there are forces which emanate, as it were, from the market mechanism itself. This anonymous power, referred to in classical terms as 'market discipline', forces all market actors into line. The fear that other actors might defeat you in this ongoing race (and that they may even be on the verge of doing so!) is a major impulse behind this type of force.

Economists differ in their opinions about the extent to which these two forms of power actually exist in concrete markets. They also dispute their desirability, in particular that of the first type. A rule of thumb in these discussions is that the larger the distance between an author and the neoclassical tradition, the greater his awareness of the first type of force, the greater his acceptance of it and the greater the likelihood that he will see certain advantages in the presence of this kind of market force, in spite of all the disadvantages. Galbraith (1952) for example points to the fact that market power reduces business risk and as such can have a positive effect on the quest for innovation. On the desirability of the second type of force there is broad consensus among economists. In standard markets and in the eyes of the majority of economists, market discipline should be severe. Hayek expresses this beautifully.

> *'Man in a complex society ... - in which the effects of anybody's action reach far beyond his possible range of vision - ... can have no choice but between adjusting himself to what to him must be the blind forces of the social process and obeying the orders of a superior. So long as he knows only the hard discipline of the market, he may well think the direction by some other intelligent human brain preferable; but, when he tries it, he soon discovers that the former still leaves him at least some choice, while the latter leaves him none, and that it is better to have a choice between several unpleasant alternatives than being coerced into one.'* (Hayek, 1948: 24).

II. Competition

In addition to exchange, rivalry or competition is an important way of coordinating action in the market, according to non-mainstream economists. Competition may be defined as the situation in which various agents pursue the same goal (De Jong. 1985). The discipline of the market depends on rivalry to a great extent. It is competition and the fear of it that makes actors continually strive towards innovation, increasing efficiency, maximizing profit. And of course, competition also unleashes a drive to attain power within the market, since power puts an actor in a position to defuse competition.

De Jong (1985: 6-15) is right to state that exchange and competition do not entirely rule out the existence of yet more ways in which market activities are coordinated. Market actors can and are permitted to cooperate in some ways. Domination of one actor by another is also possible. However, it is true to say that these last two mechanisms may never become so prominent that they come to stand in the way of exchange or competition. In that sense, exchange and rivalry remain the primary coordinators of market activity.

De Beus (1989: 173) proposes 'economic polycentrism' as a distinct characteristic of the free market. He defines this as an economy in which '*the economic power between the institutions is dispersed to such an extent that companies and families can formulate and execute their own plans independently of one another*'. Distribution of power is indeed important to a free market system. However, I see this characteristic as an implication of the presence of rivalry. After all, competition is generated to a significant degree by absence of market power.

III. Price mechanism

The third characteristic that non-mainstream economists attach to the modern free market is the presence of a monetary price mechanism. A mechanism of this kind is essential in a decentralized system to enable the coordination of actions on the basis of exchange and competition to run smoothly.

> '*Exchange can hardly become a significant method of social organization, however, if exchanges are only occasional and fortuitous. A weekly assembly for barter or the posting of public notices increases the frequency of exchange only modestly. Only with money and prices can exchange become the instrument of major instead of occasional very small-scale social organization. Prices are a device for declaring in standardized form the terms on which exchange is offered or consummated. With prices, it is no longer necessary for a person to announce tediously to each of many potential partners in exchange each of the services and commodities, together with the amounts of each, that he stands ready to take in exchange for his offer. He simply announces its price.*'
> (Lindblom, 1976: 34-35).

IV. Rational actors geared towards their own profit

Fourthly, agents in the market act in a specific *role* according to non-mainstream economists. When examining this characteristic, it is instructive to start with the neoclassical vision of this role of 'economic man', since present-day non-mainstream thinking has primarily developed as a reaction to the neoclassicists in this regard. In neoclassical theory it is assumed that people (in the perfect market) behave as economic men. Economic man is a rational being whose actions are geared towards the realization of his own interests.

The idea of economic man therefore includes specific intellectual capacities and a specific motivation. Where intellectual *capacities* are concerned, economic man is rational. Rational action should be interpreted as utilizing the available means as efficiently as possible. For economic man, the ends are a given.

Where *motivation* is concerned, economic man is focused on his own interests. In both the positive and negative sense, he operates independently of other individuals. He does not act out of solidarity, but nor does he act out of jealously. According to neoclassical tradition, this focus on his own interests does not necessarily mean that economic man is always out for his own *self-interest* (profit). A human being can define his own interest in whatever way he chooses. In this sense, Christ could even be said to have acted as an economic man when he interpreted 'dying for mankind' as his own interest.

The adequacy of this distinction between 'self-interest' and 'one's own interest' has been subject to harsh criticism from outside the neoclassical tradition. Knight (1935) and Hirschman (1992) point out that the concept of 'one's own interests' becomes tautological within the neoclassical frame of thought. If everyone by definition always acts in their own interest, then what we are left with is a meaningless conceptualization of human action. Hausman and McPherson (1993) and Sen (1977) make a similar point. They regard the idea that each man acts in his own interest as not being specific enough to constitute a proper motivational theory. This theory automatically gives rise to the need for a further distinction between various kinds of motivation to act, for example in terms of egoism or altruism. As I see it, we are also justified in saying that the neoclassical tradition confuses the fact that every person has to have a motivation for his actions with an opinion about the *nature* of this motivation. Hodgson (1988: 284) adds an empirical point to this conceptual criticism. He regards the conceptual distinction between 'self-interest' and 'one's own interest' as being of little importance in practical terms since, in practice, neoclassical economists have the tendency to equate 'one's own interests' with 'self-interest'. In the light of all these comments, it seems safe to conclude that the neoclassical tradition assumes man to be a self-interested individual, in the sense of an individual who is only interested in his own good (and that of his family) and who acts in this interest within the confines of the law and the rules of common decency.

To a certain extent non-mainstream economists also embrace the notion that we can characterize a human being *operating on the market* as an economic man. Nevertheless, the thinking of the non-mainstream economists on this point differs markedly from neoclassical thinking (see for example Weber, 1921; Knight, 1935; Hodgson, 1988; Hausman and McPherson, 1993; Hayek, 1948; Etzioni, 1989). To begin with, non-mainstream economists cast doubt on man's rationality. They see the capacities of humans in this respect as limited. In addition to this, many non-mainstream economists are of the opinion that man's rationality is not restricted to weighing up means in relation to ends. Man also has the ability and the tendency to reflect on the ends themselves. Indeed, he is more or less forced to do so, since more often than not an individual strives after conflicting goals (Etzioni, 1989). This is even true of the actor who focuses as intently as possible on economic goals: maximization of profit is just as abstract a goal as 'life' and therefore needs to be specified with the aid of often contradictory sub-goals (Lindblom, 1976:153).

Unlike neoclassical theorists, non-mainstream economists place a great deal of emphasis on the differences between real people and the idea of economic man. By nature man is not an economic creature. Many non-mainstream economists are prepared to accept that people can be described to a certain extent as economic men *when acting in the market* but only because, as they see it, the market mechanism *forces them into this role*. It is the market mechanism that brings out the economic man in people (Hayek, 1948; Schumpeter, 1934).

It must also be said that the differences between the way in which the neoclassical tradition and the non-mainstream economists approach the concept of economic man do not always lead to differences in appreciation. Many non-mainstream economists share the neoclassicists' opinion that it is good for people in the market to act or to have to act as economic men. Hayek (1948) for example makes an important link between the idea of economic man and the value of the market mechanism. According to Hayek the market forces people to be productive, innovative and efficient.

However, the neoclassical economists are the ones who go furthest in their appreciation of purely economic conduct. Within the neoclassical frame of mind, actions which are paternalistic or unselfish for example are actually undesirable on the market. Based on the assumption that each individual is most familiar with his own needs, unselfish actions are apt to result in an incorrect allocation of resources. We can therefore say that acting rationally and striving after one's own profit is not only a right within the neoclassical framework, it is also a moral duty (Baumol and Batey Blackman, 1991). Most non-mainstream economists are not prepared to go this far. In their opinion it is certainly not undesirable for motivation focused on self-interest to be mellowed by such feelings as sympathy and charity.

V. Private property

The non-mainstream economists' fifth market characteristic is private property. In neoclassical theory property is usually seen as a relationship between an actor and an object. Non-mainstream economists, sociologists and political philosophers are justifiably critical of this view (Coase, 1960; Bromley, 1991, MacPherson, 1978; Goodin, 1992). They feel that we should look upon property as a relationship between an actor, an object and the other actors with whom this actor interacts socially. Property is a social relationship. It can be described as a collection of rights which an actor or a collection of actors exercise over a certain object. Examples of such rights are the right to usufruct, the right to deny others access (exclusion), the right to buy and sell, the right of management, the right of disposition and the right of destruction.

In neoclassical thinking there is a further tendency to assume that this collection of rights is fixed and unshakeable, and always in the hands of a single individual. When an object is in ownership, neoclassicists take this to mean that one individual is in possession of all the rights which can be exercised over it. Private property is seen as the only form of property. Historically and interculturally this idea is incorrect. For one thing, the idea that certain individuals jointly hold a specific right is common in many cultures. For another, it is also customary in many cultures to take specific rights from such a collection and allocate them to specific people. For example in medieval society, it was common for one farmer to have the right to gather wood in a certain forest while another farmer had the right to keep pigs there. Even in present-day Western society, there are many situations in which the neoclassical notion that all rights are always gathered together in one and the same hand simply does not pass muster. What about such situations as rented accommodation or a listed company?

In order to gain a firmer grasp on the various kinds of property regimes, it is useful to establish a distinction between collective property, private property and objects which are not or cannot be subject to someone's ownership. The last instance occurs when there is no legal framework that makes property possible or when there are certain objects which have been explicitly placed outside the property regime. In the case of collective property various rights in the collection are distributed over several parties or several parties share certain rights. In the second case all or most of the rights are in the hands of a single individual. Nowadays, private property is by far the most widespread and familiar. The regime without property is fairly uncommon. One example can be found in Dutch law, though, in the status of wild animals which do not belong to a protected species. In the Netherlands common wild animals have no legal status.

Collective property is a very general and broad category which encompasses many variations. It includes, for example, the regime that allows everyone access to international waters. In addition to this we can consider the property of a rented house where the right of usufruct and the right of disposition are separated, or the property of a business where a particular set of rights is divided among shareholders and also between shareholders and management. Of course, common land and other forms of traditional common property also come into this category.

From a historical point of view, there is a clear and close relationship between the rise of private property and the rise of market society. Most economists and political philosophers see private property as a necessary condition for the existence and indeed the survival of a market economy. The classic argument in support of this necessity, and still the most widely used, is that actors on the market will not make an effort if they cannot be certain of reaping the rewards. Meanwhile, some argue that private property is a condition for generating adequate transactions (MacPherson, 1973: 133). When several people can exercise rights over a certain object, this is quick to obstruct the speed and the adequacy of economic dealings. It then becomes possible for one person to hold back transactions which are in the interest of many or demand a high price for his cooperation. Hayek's defence of the necessity of private property is formulated along this same line of argument. According to Hayek a market society has a great need for sharp demarcation lines between individuals. If there is no clear distinction between what is mine and what is someone else's, then it is extremely difficult for me to define my actions in terms of gearing them towards my own profit. I then become easy prey for third parties out to profit from my labour. The reverse is also true: without clear demarcation it is easy to damage someone without him being able to demand redress (Hayek, 1976: 106-112).

VI a. Freedom of trade

Freedom to act in a context in which the factors of production (capital, labour, land, organization, knowledge/intellect) are also freely tradable to a great extent, is the sixth characteristic of the market, according to non-mainstream economists. First of all, this entrepreneurial freedom means that economic actors are hindered as little as possible by rules ordained by the state. In a free market the state cannot tell a businessman how many shoes of a given colour or size he has to produce. Secondly, economic actors in the market are also free from all kinds of rules which originate in civil society. The market declares local norms and cultural quirks to be irrelevant. A contemporary example of this is the legal dismantling of the purity regulations (*Reinheitsgebote*) which applied to German beer, as a result of the liberalization of the European market.

The most important component of entrepreneurial freedom where this study is concerned, however, is that an actor on the market is permitted to disregard the public effects of his deeds. The modern market represents a sphere of limited public responsibility (see also Lindblom, 1976 or Koslowski, 1986). This limited public responsibility is of great importance to the functioning of the market. Precisely because other considerations play such an insignificant part in the market, actors can channel all their energies into maximizing their own profit on the one hand while on the other hand being utterly exposed to the discipline of the market. To put it another way, the fact that the market is a relatively independent sphere means that the logic of the market can develop as fully as possible.

> '*The market community as such is the impersonal, practical relationship* par excellence *between people. ... (This is so) because it is specifically concerned with the business issues related to the exchange of goods, and with nothing else. When the market is left to follow its own laws, it recognises only business matters and no personal matters, no fraternal and religious obligations, none of the basic interpersonal relationships rooted in the human community.*' (Weber, 1921: 383).

The legal history of Western society underlines the importance of limited responsibility for the functioning of the free market. The development of the free market also constitutes the development of a specific liability regime (Koslowski, 1986; Polanyi, 1944; Tisdell, 1993: 69).

VI b. Legitimate and illegitimate intervention in the market

The drive to achieve market freedom does not mean that the state cannot or is not permitted to impose any rules on the proceedings. The market is not a natural sphere (Polanyi, 1944; 140-141; Geelhoed, 1996: 18-19). Indeed many rules are needed if the freedom of all parties is to be guaranteed. To express it in more classical terms, freedom is only possible as freedom under the law. In a 'natural situation' ...

> '*there is no place for industry, because the fruit thereof is uncertain: and consequently no culture of the earth ... no account of time, no arts, no letters, no society and worst of all, continual fear and the danger of violent death; and the life of man, solitary, poor, nasty, brutish and short*' (Hobbes, 1651: 84).

In addition it would also be easy for market freedom to come into conflict with other liberal-democratic aims and objectives for which rules are an absolute must, such as the battle against child labour or slavery, the aim of sustainability or the right of all citizens to live with dignity. For this reason, some non-mainstream economists are constantly searching for the right balance between the importance of safeguarding the greatest possible freedom and the need for intervention on behalf of public objectives. The nature of this balance changes with time and can never be pinned down exactly. However, the non-mainstream economists do hold steadfastly to the principle that intervention may only take place *along certain lines*.

To clarify the ways in which the market may be controlled according to the non-mainstream economists, it helps to distinguish between the system level (the level of the market as a whole) and the actor level (the level of individual businesses). At system level non-mainstream economists dictate that the state can and even should provide the limiting conditions within which market dealings may occur. It is the task of the state to structure market activities. German economic law refers to this as *Dauerlenkung* (which translates roughly as control by limiting conditions) meaning that the state, by means of general rules, provides the limiting conditions to which the actors in the market are subjected (Geelhoed, 1996: 18). An important aspect of these rules is that they do not serve a specific economic purpose. They are more like 'rules of thumb' whose general effectiveness is assumed. The metaphor of the rules of a game is a fitting one here. The rules set out the limiting conditions within which the game should be played, but without specifying actions or detailing specific outcomes. Rules concerning property and liability are typical examples of these limiting conditions.

Alongside *Dauerlenkung*, German economic law sets up the concept of *Prozesslenkung* or 'process management'. Process management involves market intervention by the state in order to correct the balance with regard to specific economic objectives. Here the game analogy becomes strained since a game in which the rules are corrected as it goes along is virtually unheard of. The closest analogy we can come up with is probably that of an adult playing chess with a child and deciding to give his young opponent an advantage by not capturing a piece even though it might be possible to do so.

When in present-day Western society we talk about 'economic politics' we are often referring to process management. Industrial policy, innovation policy and energy policy are all typical examples of process management. Non-mainstream economists are not necessarily against process management at system level but do demand that great caution be exercised in this respect. Measures which interfere with market freedom have to be held in check as much as possible, as do measures which favour certain market parties to the detriment of others. The choice of instruments (legal or financial) must therefore be a careful one.

At actor level, control by limiting conditions has no meaning: at this level we are dealing with specific persons. It is impossible to think of a *general* rule that controls the actions of a specific person, without it also affecting other persons. Process management at this level is out of the question, as it would mean that an actor receives specific instructions on how to act. This does not tally at all with the principle that actors in the market should be free.

The distinction between process management and control by limiting conditions provides us with an insight into the non-mainstream economists' thinking on controlling the market. Nevertheless there is a drawback to using this pair of concepts: in practice, the dividing line between the two is not always easy to draw. After all, many instances of process management may take the guise of changes to the limiting conditions. Product safety requirements are a good example of this. In practice, therefore, it is sometimes difficult to determine whether a given measure should be classified as control by limiting conditions or process management. The minimum wage issue is a prime example. Those who regard the minimum wage as a fundamental right will regard policy in this area as a form of control by limiting conditions. Whoever does not share this view may interpret it as a form of process management and reject it accordingly.

VII. Limited responsibility
A last but important characteristic which non-mainstream economists attribute to the market is that actors only have limited responsibility for public issues. This proposition can be interpreted as an empirical claim about the actual motivation of actors on the market. In that case the proposition entails that actors on the market do not want to take responsibility for public issues. The proposition can also be

interpreted as a stipulative one about the proper working of the free market, in which case it means that markets are places where actors ought not to be burdened with responsibility for public issues. Non-mainstream economists endorse both interpretations of the proposition. As a consequence non-mainstream economists tend to claim that the market must be controlled by the state, with the state placing limits on the way in which the market operates. This implies that non-mainstream economists are proponents of the indirect responsibility model.

This conclusion seems to call for a number of qualifications. To begin with, non-mainstream economists do not *necessarily* endorse the indirect responsibility model. After all, non-mainstream economists react against the neoclassical position that economic actors are perfectly rational and completely self-interested. Non-mainstream economists hold that real people operating in actual markets deviate from the neoclassical standard. This implies that a non-mainstream economist need not deny that real people in actual markets might be willing and able to take some responsibility for public issues. By the same token, it can be inferred that non-mainstream economists need not be fanatical in their position. They are likely to endorse the claim that there are occasions on which some actors in the market take some public responsibility. However, as a rule, non-mainstream economists do not emphasize that people take public responsibility on the market, nor do they stress the desirability of such conduct. This is why it would be mistaken to see them as advocates of the direct responsibility model in some way.

At this point I would once again like to stress that 'limited responsibility for public issues' is something completely different from 'general amorality'. Some non-mainstream economists and political philosophers assert that the market is a sphere in which actors are amorally motivated and that it thrives best when people act thus. Habermas (1981, Band II, 256) expresses this notion when he refers to the market as an example of *norm-free sociality*. This concept resonates with the echo of Weber (1921: 383) who wrote: '*In the free market, i.e. the market unfettered by ethical norms, with its exploitation of the constellation of interests ... all ethical considerations among brothers are held to be rejected*'.

Most non-mainstream economists, sociologists and political philosophers who write on this subject regard this position as too extreme, both as a vision of the actual motivation of people in the market and as a stipulative proposition on the characteristics of a well functioning market. Authors like Knight (1935b), Polanyi (1944), Hayek (1948), Sen (1967), Arrow (1973), Gauthier (1986), Etzioni, (1989) Hirschman (1992), Van Luijk (1993) claim that in the real world market actors do not generally behave amorally or even immorally. Most agents in the market stick to commonly accepted conventions relating to such values as honesty, fairness, charity and sympathy.

According to the authors mentioned above, this empirical state of affairs is anything but coincidental, nor is it the product of a market that functions inadequately. To begin with, norms are needed in the market to prevent all hell breaking loose. The presence of norms and morals are a precondition for the market's existence. It is vital to ensure that Weber's 'peaceful conflict' remains peaceful. Another closely related point is that the authors who describe the market as an amoral sphere, like Habermas, often endorse the need for the continued existence of norms and values for society as a whole. Without norms and values, society would become unstable and would quickly collapse. However, for authors like Habermas these norms and values are primarily found in civil society. This position leaves itself open to the attack that in reality the market and civil society cannot be so strictly separated. When people act in the market, their role as economic man is limited to a greater or lesser extent by the other roles that they take on under other circumstances (Jeurissen, 2000).

Another point made by these authors is that norms are often instrumental to the effective functioning of the market. Mutual trust, for example, ensures that an inordinate amount of time and money does not have to be invested in checking one's business associates' every move, while fairness is another important principle without which transactions could only take place by virtue of interminable contracts covering every possible eventuality.

Needless to say, the non-mainstream economists who reject the idea of an amoral market are divided among themselves on many points. An important bone of contention is the influence of market dealings on morality. A number of authors take a pessimistic view of this matter. They feel that the market has a negative effect on morality. The market strangles as it were the conditions for its own survival (Polanyi, 1944; Schumpeter, 1944). Authors who adopt this stance generally support their position with the idea that, while civil society forms the most important source of morality, it is coming under increasing pressure as the market continues to advance. Other authors are less worried about the situation. They feel that morality is not entirely external to the market since the market also stimulates moral awareness. As Hayek (1948) states, reputation is still the greatest competitive advantage that a company can have.

On the basis of these arguments I conclude that one should not confuse 'limited responsibility for public issues' with 'limited responsibility in general'. The proposition on public responsibility only deals with an aspect of the morality of the market. The norms and values that according to most authors are present on the market, are described by Van Luijk (2001) as 'transaction ethics'. This kind of morality refers to the ways in which individuals ought to approach one another on the market. As a consequence they mainly deal with norms and values like respect, honesty and decency.

Value and success of the market

When attempting to obtain a sound understanding of the market, it is essential to get to grips with the prevailing thoughts pertaining to its value and function. Of course, non-mainstream economists value the market for its achievements in the area of economics. According to these economists, the market is the best answer to the inevitable scarcity that occurs in human society. In the words of Hayek:

> *'The market fulfils its functions better than other mechanisms known to history: historically speaking, the market mechanism facilitates more profitable production than other mechanisms and is more easily led by the wishes of the consumer. In addition to*

this, the market mechanism produces more efficiently and effectively than its historical opposite numbers and delivers its products to the consumer more cheaply.' (Hayek, 1976: part III, 74).

In order to arrive at a proper evaluation, Hayek says we should focus primarily on the long term. He regards the criterion of short-term efficiency as being of little interest.

A typical characteristic of non-mainstream economists is that they do not necessarily regard the economic significance of the market as being its most important value. The value of the market as an institution is also largely attributable to other normative factors. In this regard the non-mainstream economists focus primarily on the importance of the market as an environment or precondition for the good liberal-democratic life. The market is the basic institution where freedom can be experienced and enjoyed, and where people may achieve autonomy (Sen, 1985; Hayek, 1976).

In addition to this, many non-mainstream economists see the market as being valuable in a strategic normative sense. The market provides an important counterweight to the increasing might of the state. A strong market guarantees the desired plurality of the liberal-democratic society (Preston and Post, 1975; De Beus, 1989). Both the power of the market actors and that of the market mechanism are important in this respect. Both forms of power rein in the activities of the state. It is on the basis of these normative considerations that many a non-mainstream economist regards the order of the market as inherently right and just. Disappointing economic performance should not therefore be an automatic reason to change the way the market works. The market does not only owe its existence to its economic functionality.

One last aspect of the market which deserves to be examined briefly is its success. Many economists have wondered why the market is so successful in comparison with other economic systems. Non-mainstream economists have come up with a number of answers. I regard them here as supplementary to one another. Adam Smith explains the success of the market system on the basis of division of labour, which in his view flowed from the *'propensity to truck, barter and exchange'*. This division of labour facilitates an enormous rise in the productivity of labour because it opens the door to the development of specialization (*'dexterity'*), time savings and mechanization (Smith, 1776). Weber (1921: 382-385), meanwhile, links the success of the market to the purely businesslike and rational way in which actors in the market relate to one another. In their pursuit of profit they are not hampered by capricious external forces or cultural obligations. The market is no respecter of rank, so that the best people rise to the top in a continuous cycle. Hayek (1976) Dahrendorf (1966) and Lindblom (1976) mainly contrast the market with the hierarchical planned system. To a large extent, they put the runaway success of the market down to its decentralized character. This makes the market far more effective in terms of getting the right information to the right person at the right time. Planned systems are much more cumbersome when it comes to processing information. In addition to this, Hayek also emphasizes the powerful feedback mechanisms which are present in the market system. In the market every individual reaps the rewards of his own success. Schumpeter (1944) emphasizes the innovative character of the market system. The competitive nature of the market, along with its other characteristics means that actors are on a constant quest for innovation, thereby creating opportunities for entrepreneurs.

2.3 MARKET DESCRIPTION II: NEOCLASSICAL MARKET CONCEPTS

The neoclassicists describe the market using the model-based approach that is so characteristic of their tradition. At the heart of this approach lies the idea of a market that functions perfectly. This model situation is sometimes described as one of full competition (Samuelson and Nordhaus, 1985: 46). I side with writers such as Baumol and Batey Blackman (1991) in preferring the term 'perfect market'. In this context a perfect market can be described as a market which has reached maximum efficiency. The factors of production cannot be so rearranged that one party can be better off without affecting another party negatively. Nowadays neoclassical economists often interpret this 'being better off' in terms of utility or satisfying revealed preferences. They speak of an *economic optimum*. Since many a neoclassical economist would argue that all values can be converted into economic values (see Section 2.1), this situation is also referred to as a *social optimum* (Tisdell, 1993: 7).

Neoclassical economists work from the assumption that the market system as it actually exists should aim to equal this model situation. The 'true' free market system can therefore be described as a system which corresponds with the characteristics of a perfect market.

Market characteristics described in model terms

The features which the neoclassicists attribute to the perfect market show many similarities to those of the historical approach. Only a few characteristics are added. To begin with, all parties in the perfect market have all the relevant information at their disposal. In terms of information, the market is transparent. This would indeed appear to be a necessary condition for attaining maximum efficiency. In addition, neoclassical economists nowadays assert with increasing regularity that activities in the perfect market should be free of transaction costs. In the perfect market, obtaining information, weighing up options, negotiating and securing rights does not cost the actors anything in terms of money, effort or time. This assumption concerning the absence of transaction costs on the perfect market was developed by the non-mainstream economist Ronald Coase (1937). He made it clear that in reality there are costs associated with '*the use of the price system*'. This idea concerning transaction costs - Coase referred to 'marketing costs' - was later elaborated on by other non-mainstream economists and was subsequently adopted by the neoclassical tradition. Dahlman described the concept of transaction costs most specifically as '*search and information costs, bargaining and decision costs, policing and enforcement costs*' (see Coase, 1988: 6 and Hodgson, 1988: 200).

Yet these supplementary features are not all that sets the neoclassicists apart from the historical market concept. They have a general tendency to adopt the characteristics we categorized as historical and take them a step or two further. For example the neoclassicists are not prepared to settle for a healthy spirit of commercial competition. Their perfect market must be governed by 'full competition'. As a rule, this situation is currently typified as one in which '*nobody has control over prices*'. '*Perfect competition exists only when no farmer, business, or laborer is a big enough part of the total market to have any personal influence on market price*' (Samuelson, 1985: 46, 503-504). We can also say that in the situation of perfect competition no one has any power in the market. Everyone is subjected with equal force to the discipline of the market.

The neoclassicists also accentuate the assumption of rational market behaviour. According to the neoclassical model, market actors act perfectly rationally. They are entirely focused on their own profit and weigh up the options meticulously at every turn in order to assess which provides the greatest advantage, however marginal it may be. The model endows actors with so much intelligence that they never err in their calculations.

Another position which the neoclassicists take a step further is the argument that private property is very important to the free market. In the perfect market *all* economically relevant factors of production are in private hands. This position is often regarded as synonymous with the idea that in the perfect market all goods are adequately priced. From the neoclassical perspective it is not difficult to understand this equivalence. A perfectly rational owner will try to obtain the greatest possible benefit from his possessions and will therefore price his goods. His perfectly rational nature means that he will succeed in doing so as rationally as possible.

This brings us neatly to the next neoclassical accentuation. For the neoclassicists, it is not enough for the price mechanism to function reasonably well. In the model situation, the price mechanism too works perfectly. All economic values are brought into relation with one another in terms of price and weighed up against each other. This view not only imposes greater demands, it also bestows an additional function on the price mechanism. It is no longer just a mechanism that provides information on the supply and demand of economic goods, but it also becomes a method of comparing the relative value of various goods.

External effects in focus

The argument that the price mechanism works perfectly in the model situation implies that there are no external effects in the model situation. Given the great relevance of the 'external effects' concept for environmental economists, I will now explore this area in greater depth.

An external effect is the effect of an action by A or the effect of a transaction between A and B, which is of economic significance for C, but for which no price can be stipulated or payment demanded. For this reason, this effect is not calculated in the price of the goods in question. When C benefits from this effect, we refer to a positive external effect: if this effect places C at an economic disadvantage, then we speak of a negative external effect.

If we assume that the environment is an economic good, then we can typify environmental problems as examples of negative external effects. If a factory located upstream releases toxic chemical waste into the river, polluting the water and killing fish, this will have economic effects on a fishing village further downstream. The people there will be forced to pay for the purification of their drinking water supply. What is more, part of their income will disappear. If the villagers are unable to make the factory pay for these expenses, this can be termed a negative external effect. An example of a positive external effect would be an ice cream vendor who sells ice cream on a warm summer's day at the edge of the forest. The forest owner provides the vendor with extra custom because his forest attracts more people to that location than would otherwise come. If the owner is unable to make the ice cream vendor pay for this service, this can be termed a positive external effect. The vendor receives benefits for which he does not have to pay.

In abstract terms, the consequence of external effects is always the same: allocation is disrupted, which means that maximum efficiency cannot be achieved. In this respect positive external effects are just as undesirable as negative external effects. When the production of a commodity results in a negative external effect, this means the commodity is being produced at too low a price. As a consequence, *too much* of that commodity is produced. In the case of a positive external effect, the production of a commodity has benefits for which the producer is unable to exact payment. The benefits leak away, as it were. Accordingly, the goods which bring about the positive external effects are produced *in insufficient quantities*.

Two common misunderstandings surrounding external effects need to be refuted. To begin with, it is not necessarily the case that the party who causes an external effect benefits from it in terms of increased profits. In a competitive market, a factory that pollutes a river may have been forced to lower its prices to such an extent that any cost-cutting advantage is entirely passed on to the buyer. Nor can external effects be seen as objectively observable facts. In an interdependent economy, not every effect that one actor has on another can be classified or is classified as an external effect. Classification as an external effect is preceded by a normative assessment. Mishan (1981: 385-395) clarifies this point by giving two examples of effects which we do not classify as external. One is that of a man who loses out in terms of *welfare* because he is jealous of the car his neighbour has purchased. Another example is a product that goes up in price due to the increase in demand caused by new buyers. This has a negative effect on the old buyers, but they cannot complain of external effects.

A functional definition of the market

The model-based approach is not the only one that we encounter within neoclassical economic theory. In addition to this conceptualization, the market is defined in functional terms. A peculiar aspect of this functional definition is that it usually remains *implicit* within neoclassical theory. Stated explicitly, the functional conceptualization describes the market as the sphere in which *private interest* meets *public interest*. The neoclassical authors who state this most outspokenly include Freeman et. al, (1973) and Pigou (1920). Hirschman (1992: 45) refers critically to this approach when he speaks of the market as the sphere of social harmony. Each individual in the market operates in accordance with the dictates of his own interests, but the disciplinary forces of the market make the sum of self-interested transactions equal to the greatest good for all (Samuelson and Nordhaus, 1985: 46).

This supposed ability of the market to shape the sum of all self-interested endeavour into the public good may rightfully be called an astonishing quality. C.L. Schultze (1977: 18) even goes so far as to say: *'(h)arnassing the "base" motive of material self-interest to promote the common good is perhaps* the *most important social invention mankind has yet achieved'*. If the market is indeed capable of bringing about such harmony, then he can hardly be accused of overstatement. What more can a society wish for than a situation in which the common good is served by everyone following their own interests? This resolves the fundamental quandary of how to pluck the fruits of cooperation and interdependence, while preventing the disadvantages of conflict.

Adam Smith is often regarded as the man who discovered the miraculous operation of the market. He used the appealing metaphor of the 'invisible hand'.

'As every individual, therefore, endeavours as much as he can both to employ his capital ... every individual necessarily labours to render the annual revenue of the society as great as he can. He generally, indeed, neither intends to promote the public interest, nor knows how much he is promoting it. ... (H)e intends only his own gain, and he is in this, as in many others cases, led *by an invisible hand to promote an end which was no part of his intention. Nor is it always the worse for the society that it was no part of it. By pursuing his own interest he frequently promotes that of society more effectually than when he really intends to promote it.'* (Smith, 1776: 400).

In this context it is important to explore the exact meaning of Smith's words. By stating that there is an invisible hand at work in the market, Smith meant that the market realizes its social function *behind the people's backs*. The power of the market lies in its ability to utilize the unintended consequences of actions. The adequate allocation of goods for all takes place while each individual is only working to achieve an optimum allocation for himself.

Present-day neoclassical authors regularly refer to Smith or his famous metaphor when explaining the working of the market. One might question whether it is always appropriate for them to quote Smith as evidence for their own theory. He is more modest about the functionality of the market than his 20th century disciples, and does not speak of the market as achieving an optimum situation. For Smith and his contemporaries it was revelatory enough that the market, a sphere where the medieval Christian ethics of brotherly love and altruism were so derided, ultimately produced a great many positive effects.

Value and success of the market

From a neoclassical perspective, the main function of the market is economic. This is also the main reason why neoclassicists value the market. The neoclassical view may therefore be called an instrumental one. The neoclassicists do not attribute any intrinsic worth to the market but they do value it highly as an instrument for economic allocation. The market gives the best answer to the four fundamental economic questions, as posed by Samuelson (1985): what to produce, how much to produce, for whom to produce and how to produce? The market is the best answer to scarcity. In support of their arguments the neoclassicists do not look to history, as the non-mainstream economists sometimes do. Instead they are convinced that they can use a logical-mathematical model to prove the market's status as the most effective economic system. Under perfect market conditions, equilibrium will be achieved at maximum production. Here once again we encounter a striking difference with the non-mainstream economists: while the neoclassicists resort to equilibrium to prove the success and value of the market, non-mainstream economists often see the market's dynamic nature as its main strength.

The neoclassical vision of public responsibility

The many differences between the neoclassical tradition and non-mainstream thinking imply that the two approaches must differ in their view of public responsibility in the market. As before, the neoclassicists take a more extreme stand. As they see it, acting out of self-interest is not only rational or to be expected in the market; neoclassical theory looks upon it as a *duty* (Baumol, 1975 and 1991). This duty follows from the functional legitimization of the market within neoclassical thought. If the legitimacy of the market fully depends upon its economic performance and if the economic optimum is only reached when all actors are motivated by self-interested, then acting self-interestedly arguably becomes a duty. All this means that the indirect responsibility model is an inevitable aspect of neoclassical thought.

The relationship between the market concepts

My account of the three market concepts (empirical-historical, model-based and functional) is now complete. Since I intend to concentrate on an analysis of the thinking of neoclassical environmental economists in this chapter, it is my primary concern to explore the second and the third market concepts. By way of context, however, I will begin with a brief comparison of the first and second market concepts. The model-based description appears to be nothing more than an accentuation of the historical market concept. However, this would be to underestimate the full implications of the differences between the two. Hayek and other supporters of the historical market concept are of the opinion that, theoretically speaking, neoclassicists understand little about economic processes and that their theoretical mishits mean that they repeatedly arrive at practical or policy-related recommendations which are incorrect.

Hayek maintains for example that the requirement of complete information in the market is a foolish one. The information available will always be limited, if only for the reason that the human capacity for dealing with information is severely limited. One of the most important economic questions is therefore: how can we best deal with our limitations in this field? The market is the answer to this question. Accordingly Hayek reasons that the person who sets the requirement for complete information in the market has failed to grasp why a market is actually useful and necessary in the first place. Hayek is also of the opinion (1948: 11) along with many others including Coase (1988: 3-4) that it is nonsense to view man as an actor who rationally pursues clearly defined targets. '*The rational utility maximizer of economic theory bears no resemblance to the man on the Clapham bus, or, indeed, to any man (or woman) on any bus.*' (see for example also Nelson and Winter, 1982). According to Hayek the assumption of perfect rationality also betrays the limitations of the neoclassicists' theoretical capabilities. We need a market for the very reason that man has a tendency to be lazy and irrational. The market filters out stupidity and forces people to be productive.

In more general terms, Hayek (1948: 94) and others criticize neoclassical thinking for putting the cart before the horse time and time again. '*(T)he modern theory of competitive equilibrium* assumes *the situation to exist which a true explanation ought to account for ..*'. One example to illustrate this criticism concerns the requirement of perfect competition. According to the neoclassicists perfect competition is achieved when no one is still capable of determining the price of a product. Hayek sees this definition of perfect competition as absurd. When no one is still able to make a difference, it will spell the end of competition. The requirement of full competition is defined in terms which are contradictory.

Galbraith (1991) adds to this criticism the consideration that the neoclassical conceptualization of scarcity is incorrect in the present-day context. Neoclassicists interpret present-day scarcity as absolute scarcity. Where they are concerned there is little conceptual difference between today's economy and the subsistence economy of yore. According to Galbraith modern affluent society cannot be compared to such a society from the distant past. Scarcity has become relative scarcity. Most people nowadays do not have to pinch pennies before deciding what to spend them on. This has changed the way people make choices and it follows that thinking about economic processes should change too.

These differences in position between neoclassical and non-mainstream economists are not only interesting from an academic viewpoint. In terms of practice and policy, people arrive at different recommendations on the basis of different theories. For example economists like Hayek and (J.M.) Buchanan object to obligatory quality standards for products.

2.4 THE NEOCLASSICAL DIAGNOSIS AND THERAPY

How do neoclassical environmental economists assess the problems related to the realization of sustainable development? From their point of view, how can the market be controlled in this regard? What is their diagnosis of the situation and what is their therapy? By answering these questions, I hope to move closer to understanding the neoclassical mental model of social organization.

The concept of 'market imperfection' provides a key to neoclassical thinking about environmental problems. Neoclassical environmental economists are sometimes remarkably relaxed in their descriptions of this crucial concept. A market imperfection is defined as a real-life situation in which the market does not attain the social optimum. An example of this approach can be found in Clem Tisdell's description:

> '*While the market system is a powerful, relatively inexpensive and responsive mechanism for allocating resources, in practice it is less than perfect. Market failure occurs. Market failure is said to occur when the price mechanism or the market system, the so-called invisible hand, fails to bring about the social optimum.*' (Tisdell, 1993: 7).

In my opinion it would be wise to define the concept of market imperfection more clearly. The market can obstruct the 'social optimum' in three ways. Given the analytical and political-administrative differences between these obstructions, we must be careful not to lump them together. With this in mind, I wish to introduce the distinction between market imperfections (in the narrow sense), market disadvantages and market restrictions.

I define *market imperfection - in the narrow sense* - as a situation in which the empirical or *de facto* existing market does not meet the conditions imposed on the perfect model market. In practice, this results in a failure to achieve the economic or social optimum. A situation in which the consumer has to pay too high a price for a product due to monopoly formation is a typical example of a market imperfection in this limited sense. Since the empirical reality does not meet all the criteria for a perfect market - in this case the demand for 'full competition' - market forces are disrupted. The market as it actually exists does not succeed in achieving what it achieves in the model situation.

It is important to separate this narrower description of market imperfection from a *disadvantage* or *undesirable situation* which occurs in a society where the market has been perfectly formed. One example of this would be excessive social inequality (Knight, 1935; Eijgelshoven and Nentjes, 1993: 13). In the perfect market, there is an ever-present danger that social inequality could take on undesirable proportions. One reason for this is the difference between individuals in terms of their natural and social properties. This in turn influences their economic relevance and/or their position in the interplay between supply and demand. I will refer to these socially undesirable situations which occur specifically as a result of a perfectly designed market as *market disadvantages*. The reasons for distinguishing these from market imperfections should be clear. When thinking about how to relieve or reverse a market disadvantage we cannot resort to the theory of the perfect market. In order to deal with market disadvantages, we have to rely on the state, civil society or an imperfect market design.

The concept of *market restriction* refers to a situation in which there are goods, services or objectives which *by their very nature* fall *outside of the reach* of the market. The market is incapable of producing such items, even more so if it is perfectly organized. Goods which liberal-democratic society feels should not be produced by the market also belong to this category. The question of exactly which goods fall outside the reach of the market, perfect or otherwise, continues to divide academics. It is a difficult question to answer in any case as it depends partly on historical factors, such as the degree of technological development. Within the liberal-democratic tradition, however, there is a reasonable measure of consensus regarding the notion that the market is not suited to determining such processes as socialization and upbringing. In a similar vein, few believe that the market can adequately guarantee objectives like safety, social justice or a proper legal system.

The diagnosis

Neoclassical environmental economists diagnose environmental problems as market imperfections in the strict sense of the term as outlined above. The argumentation underlying this conclusion can be summed up in five surprisingly simple steps.

1. The environment is an economic good.
2. The perfect market achieves an optimum allocation of economic goods.
3. Modern society is racked by environmental problems.
4. With regard to environmental problems, the allocation of factors of production does not therefore lead to the optimum allocation.
5. Environmental problems constitute a market imperfection.

So neoclassical environmental economists explain environmental problems in terms of an inadequately functioning market. One point which should be mentioned here is that neoclassicists do not regard the mere existence of processes like the extinction of plant and animal species, depletion of water resources or pollution as economically relevant environmental problems. Such processes only attain this status when their occurrence leads to disruptions in economic allocation. As long as they do not lead to a *loss of welfare*, there is no economic problem and therefore no need to control the market. In a neoclassical paper we are therefore unlikely to read that 'the environmental problems should be resolved'. Instead neoclassicists recommend that society go in search of the 'optimum pollution point' (Pearce and Turner, 1990: 60-101).

Taking the analysis one step further, we may ask why the working of the market is hampered by structural imperfections when it comes to the environment? Neoclassical environmental economists often account for the market imperfections associated with environmental problems as a consequence of either external effects or the collective character of many environmental goods (Freeman, Kneese and Haveman, 1973: 71-79). A concise and oft quoted definition of a collective good was given in 1954 by Samuelson. He defined a good '*to be collective if a person A's consumption of it did not interfere with a person B's consumption of it'* (see Mishan, 1981: 431). This description of a collective good does not seem fitting in the context of this study. For example it leaves no place at all for collective goods whose production is obstructed by *jointness of production*. Still I do not wish to delve all too deeply into the definition of the concept of collective goods at this juncture. The reason for this decision lies in Mishan's assertion (1981) that externalities and collective goods are closely related analytically. Economic problems which are conceptualized in terms of 'collective goods' can be integrally translated in terms of 'external effects'. I go along with Mishan in this train of thought and will therefore refrain from a separate exploration of the theory of collective goods as a way to analyse the origin of environmental problems.

The neoclassical environmental economists have almost rounded off their diagnosis when they arrive at the notion that environmental problems can be conceptualized as 'externalities'. However, it remains to be explained why externalities relevant to the environment are so plentiful in the market as it actually exists, when they are entirely absent from the perfect model market. The general explanation given by the environmental economists is that the environment has no price in reality. This lack of price means that people take no account of the environment when they make their decisions, thus giving rise to externalities (Pearce, Turner and Bateman, 1994: 72-73; Van Ierland, 1990: 3; Barbier et al., 1997).

The next obvious question is why the environment has no price in reality. Neoclassical environmental economists attempt to explain this in various ways. The most common explanation is that property relationships have not been adequately defined in reality. As Herfindahl and Kneese (1974: 51) state *'But with respect to externalities, the central problem is "common property"'*. The absence of a clearly defined and enforceable property regime applying to environmental goods in practice, means they are priced inadequately or not priced at all, which gives rise to environmental problems. In short, the environment lacks an owner. As a result, it is neither managed, protected nor capitalized on. It is important to realize in this respect that most neoclassical economists implicitly or explicitly regard private property as the only adequate property regime. In terms of neoclassical thinking therefore 'the lack of an adequate property regime' should be translated as 'the lack of private property'. The neoclassical economists therefore throw collective regimes and regimes without property together in a single category. We find one example of this in the above definition by Herfindahl and Kneese. These authors cast the problems in terms of 'common property' where it would have been more appropriate to speak of a regime without property. To avoid complicating the matter still further I will refrain from discussing this false reduction of common property regimes to regimes without property. (For criticism of this position, see Bromley, 1991; McCay and Acheson, 1990; McCay and Jentoft, 1996).

The therapy

In the previous century, liberal democracy mainly dealt with externalities by screening or *removing* goods that are sensitive to these forces from the influence of the market. The state made rules to prohibit the causing of externalities, linked these to sanctions and attempted to enforce the rules by means of a civil service apparatus of inspectors and policemen. This form of administration is known nowadays as 'management by command' (Bovens, 1996c), '*command and control*' (Schelling, 1984: 27; Pearce and Turner, 1990: 161) or '*hierarchical governing*' (Kooiman, 1999; 6; Van Vliet, 1992: 13). Many feel that this form of administration is now in need of review. Command and control is thought to be ineffective and inefficient, resulting in tutelage, implementation problems, the piling up of regulations, and the overload of the state (for an extensive discussion of these issues see Chapter 3).

Within the context of this discussion on the ways in which the market can be controlled, the neoclassical environmental economists formulate an alternative. In a nutshell, they feel that market imperfections should be repaired by adapting the real-life situation to match the model. The solution supported by the environmental economists can also be described as *marketization*. Externalities should be brought into the market not removed from it. Although neoclassical environmental economists do not often express it in such terms, we can also say that getting rid of market imperfections constitutes an extension of the market domain. An area of social intercourse that was first coordinated by another fundamental institution is now brought within the market's sphere of influence. A term that neoclassical environmental economists themselves often use to denote their therapy is 'pricing'. A price tag needs to be attached to unpriced environmental goods (Schelling, 1983: ix; Pearce and Turner, 1990: 84). In Section 2.1, I referred to the proposed strategy of the neoclassical environmental economists as 'market-based regulation'. This is the concept I will adhere to here.

Neoclassical environmental economists put forward various ways in which market-based regulation or pricing can take place. I will distinguish between two main paths: market-based regulation governed by public law and market-based regulation governed by private law. In regulation by public law, the state corrects market failure by means of charges and taxes. Environmental goods are invested with economic significance because the state attaches a price to them and enforces this within society. An example of this approach would be charges on fossil fuels to combat the emission of nitrogen and carbon dioxide.

Market-based regulation governed by private law means that the state changes the institutional structure of the market in such a way that specific private parties (agents) are given an interest in protecting the environment and the opportunity to defend this interest by law (see Tisdell, 1993). In most cases this will involve changes to the legal structure of the market (and/or society). Examples include the allocation of property rights or changes to liability legislation. So, the big difference between market-based regulation based on public law and its private law equivalent is that in the public alternative, the state is the actor that allocates a price, thereby protecting the environment. In the private alternative, meanwhile, the structure of the market is changed in such a way that it becomes possible for private agents to put a price on the environment. In the case of a public law system, differences can be settled by the administrative courts where necessary. After all, it is the state that sets the prices. When market-based regulation is governed by private law, a dispute regarding the price of an environmental good becomes a legal matter between two private parties, and it is therefore up to the civil courts to decide the case.

When it comes to evaluating the public and private options, most environmental economists are of the opinion that *prima facie* private law strategies are preferable (see for example Freeman et. al, 1973; Pearse, 1994). After all, many environmental economists see the lack of an adequate property regime as the basic reason for the occurrence of externalities pertaining to the environment. Privatization is seen as a solution that tackles this problem at its source: the environment is given an owner who is caught between the pursuit of profit and the discipline of the market and who will therefore make sure that the environment is equipped with the right price. According to the environmental economists, privately governed market-based regulation will also keep the role of the state to a minimum. All the state has to do is set up the new institutional rules and then it can remain in the background. Furthermore in the private alternative, the price of environmental goods is actually determined on the market. It is not a surrogate price imposed by the state.

Nonetheless, many economists take a cautious approach and have a tendency to qualify their arguments in favour of a private law strategy. In practical terms, the scope of such an approach is limited.

> 'There are two main barriers to greater reliance on the vesting of property rights and private exchange. The first is the size of environmental units to which property rights must be vested. An air shed or a major river system are inherently indivisible. To secure the economic benefits stemming from the creation of property rights, control over the whole system must be granted to a single entity. Yet control over large environmental resources would convey enormous economic power to the owner. The result would be monopoly power and the associated misallocation of resources. The second barrier is the presence of public good attributes in many types of environmental services. The inability to exclude those users who had not paid for the service would make it ... impossible for the owner of an environmental resource to collect (his) resources.. .'
> (Freeman et. al, 1973: 96).

A short evaluation

As a stepping stone on the way to analysing the neoclassical environmental economists' mental model of social organization, I would now like to offer a brief evaluation of their diagnosis and therapy. This evaluation should primarily be seen as a prelude to the analysis of the environmental economic mental model that follows. A striking aspect of neoclassical environmental economic theory is that ultimately it is difficult to pin down the central claim of this thinking. As I see it, the theory jumps back and forth between two assertions. The first claim, and a very moderate one at that, is that the state should learn to apply instruments that are in keeping with the workings of the market. The second, more ambitious claim is that market-based regulation can serve as a fully fledged alternative to management by command.

In my view the neoclassical environmental economists have rallied round the second claim, which is often reflected in the work of these economists. For example when they express the hope that *we* (society) *have no need for government regulation ... (if) the market will take care of itself* (Pearce and Turner, 1990: 70). In addition to this, the state and government intervention are often set up in opposition to the market and market-based regulation. However, if this strategy only represents a change to the way in which the state should manage the market, it is unnecessary and unjustified to draw this dividing line. Market and state would then be two sides of the same coin, with the state creating the conditions under which the market can function adequately.

Another remarkable aspect of present-day neoclassical environmental theory is that it is formulated in relative isolation from the general debate about public policy and the way modern society should be governed. While the general debate loudly laments the passing of our ability to shape society, serenity and peace reign in the neoclassical camp. At no point do the neoclassical environmental economists question whether society has the power to restructure the market in the desired direction. This can be regarded as curious since the theory on market-based regulation requires making considerable changes to the structure of the market.

In a way, this position as prime advocate of Progress is admirable and justified, since present-day discussions all too easily resort to the argument that it is impossible to shape society. However, this is hardly an enviable position to adopt if there is no explanation as to why society is free to ignore evidence from all quarters regarding the unruly character of present-day reality. Words like naive spring to mind. The painful aspect of the situation is that neoclassical environmental economists do not in fact have an explanation for this state of affairs. They restrict themselves to the observation that politicians and society pay but scant heed to environmental economic analysis (Pearce and Turner, 1990: 96-98, 159-169).

The diagnosis and therapy proposed by the neoclassical environmental economists has not remained unexamined in other academic fields. In addition to praise and shows of support (Bressers, 1983) we also encounter a number of standard criticisms. Neoclassical environmental economists are accused of seriously underestimating the problems involved in implementing their own instruments (Andersen, 1994: 25-30). Given the general debate on the bankruptcy of management by command, such criticism is not to be taken lightly. Another oft heard objection is that these economic instruments are not exactly the most elegant tools conceivable in terms of legal certainty and equality before the law. The critics also feel that the deployment of economic instruments offers no certainty as to the effectiveness of policy. Those active in the market still have the freedom to ignore price stimuli or to simply pass these on to the customer when setting their own prices. In instances where the aim is to achieve an absolute reduction, the economic approach can certainly be regarded as unsuitable (Van Vliet, 1992; Brussaard, 1991: 38-45).

The majority of these points are also recognized by critical environmental economists. On the basis of his case study, Marcus for example states that regulatory problems can be a good deal more complicated than the theory of the perfect market allows. Accordingly, he warns against being too optimistic when it comes to market-based regulation. *'The economists' idea ... overlooks compatibility and various technical and political problems that prevent its complete realization.'* (Marcus, 1982: 173). Problems relating to setting the levels of price stimuli and controls on companies are major practical stumbling blocks. Tisdell (1993) subjects various proposals put forward by prominent neoclassical economists to a critical analysis, including Baumol's *'fiat approach'* (i.e. a system of charges) and Dales' plan for the sale of tradable emission rights. His conclusion is that the success of these approaches depends upon conditions which are so stringent that in practice they would be very difficult to comply with.

The criticisms dealt with so far can be summarized by saying that economists are not fully enough aware of the translation that needs to be made between the simple world of the model and the unruly nature of the real world outside it. The *Ricardian vice*, the phenomenon described by Schumpeter (1954: 472-473) as the overeagerness of economists to put their ideas into practice, is apparently still alive and well in the sphere of economic thought.

A different critical tack and one that is of greater relevance to my purpose is the suggestion that the neoclassical environmental economists have narrowed their conceptual view too much. They seem to assume in advance (or by definition) that market imperfections underlie environmental problems. However, this one-sided view is not easy to justify in the real world where market forces are not the only ones at work. The state for example also makes its presence felt. Nevertheless, the neoclassical environmental economists seem to automatically discount the dysfunction of the state as the possible cause of at least some environmental problems. The environmental economists have a blind spot. As Andersen (1994: 12) puts it: *'The theory of government failure is less developed than the economic theory of market failure'*. Likewise, Pearce and Turner remark that (1994: 89) *'(w)hile we are all used to the idea that governments should put things right, we are less familiar with the idea that government policies ... can and often do damage the environment'*.

Ronald Coase expressed this line of criticism powerfully in 1960. According to Coase, it is not at all difficult to show *empirically* that the state is the ultimate cause of a considerable number of environmental problems.

> *'Most economists would appear to assume that the aim of governmental action in this field is to extend the scope of the law of nuisance by designating as nuisances activities which would not be recognized as such by the common law. And there can be no doubt that some statutes, for example, the Public Health Acts, have had this effect. But not all Government enactments are of this kind. The effect of much of the legislation in this area is to protect businesses from the claims of those they have harmed by their actions. There is a long list of legalized nuisances'* (Coase, 1960: 24).
> *'(Therefore, it) is my belief that economists and policy-makers generally, have tended to over-estimate the advantages which come from governmental regulation.'* (idem: 18).

However strange it may seem, according to Coase, economists either cannot or will not face this fact.

> '*Most economists seem to be unaware of all this. When they are prevented from sleeping at night by the roar of jet planes overhead (publicly authorized and perhaps publicly operated), are unable to think (or rest) in the day because of the noise and vibration from passing trains (publicly authorized and perhaps publicly operated), find it difficult to breathe because of the odour from a local sewage farm (publicly authorized and perhaps publicly operated) and are unable to escape because their driveways are blocked by a road obstruction (without any doubt, publicly devised), their nerves frayed and their mental balance disturbed, they proceed to declaim about the disadvantages of private enterprise and the need for government regulation.*' (idem: 26).

What are we to make of Coase's criticism? At first sight it seems a little misplaced. After all, neoclassical environmental economists are among those heaping criticism upon the way in which the state functions, accusing it of ineffective, inefficient, lax and contrary management (Freeman et al., 1973: 80-107; Schultze, 1977). We can even go one step further in this direction. The whole strategy surrounding market-based regulation is aimed at transferring tasks and responsibilities from the realm of the state to the market. The market has the unenviable task of taking over the work of a state that operates cumbersomely, hierarchically and bureaucratically. A second response to Coase could be that his theory is also wide of the mark in an analytical sense. Should we really take Coase seriously when he writes that economists '*proceed to declaim about the disadvantages of private enterprise and the need for government regulation*'? Isn't market-based regulation celebrating the advantages of private enterprise?

This is indeed the case, but nevertheless I feel that we cannot simply push Coase's critique aside. Under closer scrutiny, the reasoning employed by the neoclassical environmental economists does appear to be slightly warped. To begin with, they really do have a blind spot when it comes to the performance of the state. Contrary to Coase's claims, this is not so much related to the actual failures of the state in *present-day practice* but in its *potential*. Neoclassical environmental economists hardly see any boundaries or barriers when it comes to the possibilities open to the state for creating a perfect market. One aspect of this is their failure to acknowledge the scale of the administrative effort that would be necessary in order to acquire knowledge of the correct price stimuli and/or the effects or side effects of institutional changes. They also fail to sufficiently recognize that even in a perfect market there is still a genuine need for a system of implementation and enforcement (Van Vliet, 1994; Pearce and Turner, 1990: 97, 107, 160; Tisdell, 1993).

Secondly, there seems to be something very strange about the logic of neoclassical environmental theory. On the one hand, the argument for market-based regulation is grounded in the idea that the state acts inefficiently, cumbersomely etc. and that it should be relieved of some of its tasks by transferring some responsibility for the environment to the market. On the other hand, market-based regulation *presupposes* a great deal of tough state-action. After all, it will be left up to the state to restructure the market. Could the neoclassical faith in the perfect market perhaps conceal a belief in the perfect state after all? Is this perhaps the best way to interpret Coase's point?

In the sections that follow, my aim is to answer this question with reference to an analysis of the neoclassical mental model of social organization. With this in mind I have put together a brief historical excursion in the next section, to take us through the development of economic thinking as regards market and state. Without this backward glance, the thinking of the present-day neoclassical environmental economists is difficult to understand. Special attention will be given to the thinking of Arthur Pigou. Present-day neoclassical environmental economy is heavily indebted to him.

2.5. ECONOMIC MENTAL MODELS FROM LAISSEZ FAIRE TO PIGOU

Before I embark on my short historical excursion, I would like to point out that, when it comes to mental models of social organization, we should make a clear distinction between laissez-faire thinking and the neoclassical tradition. The neoclassical mental model of social organization is a reaction to and critique on laissez-faire thinking in this regard. It is worth pointing this out explicitly, as it is something that is not always clear to those describing or criticizing the neoclassical approach to socio-political thinking. They should not be particularly blamed for their confusion, since the neoclassical authors themselves contributed considerably to this lack of clarity. The neoclassical tradition made its switch to the model-based approach at the end of the 19th century. This switch meant that economic theory withdrew, as it were, from reality and focused instead on an ideal model world. As a result, many of the scientific conclusions drawn by the early neoclassicists all related to the ideal world of the model. One of these conclusions was that laissez faire would lead to an economic optimum in the model world. In their more rigorous moments, the neoclassical economists were quick to add that the real world deviated from the model and that putting laissez faire into practice was by no means a good idea in many cases. However, many neoclassical authors were regularly troubled by less insightful moments in which the distinction between the model and the real world was not emphasized enough. This sometimes created the impression that neoclassical authors were defending laissez faire, despite the fact that authoritative neoclassical authors like Pigou explicitly rejected this position (see also Keynes, 1926).

> *'Yet some of them cannot be entirely absolved from the charge of having helped to maintain faith in it* (laissez faire that is - wd) *by their support, and their apparently logical proof, of its doctrine. This is especially true of Léon Walras and his immediate disciples. ... (But) Walras realized that the theory, if it was to be maintained at all ... must be proved more satisfactorily than has hitherto been done. "Il faudrait prouver que la libre concurrence procure le maximum d'utilité". And this view was in fact the starting point of his own work in economics. It is almost tragic, however that Walras, who was usually so acute and clear-headed, imagined that he had found the rigorous proof, which he missed in contemporary defenders of the free trade dogma, merely because he clothed in a mathematical formula the very arguments which he considered insufficient when they were expressed in ordinary language.'* (Wicksell, 1934: 74).

Present-day neoclassical thinking on the relationship between the market and the state has its roots in the second half of the 19th century. Then as now a lengthy, hard-fought and highly politically charged debate was raging with regard to the proper limits of the functions and agency of governments. Economists exerted a powerful influence on the nature of this discussion (Searle, 1998). A prominent group, under the intellectual leadership of figures like Richard Cobden (1865) of the Manchester School and Mrs A. Marcet (1816), adhered to the belief that as many social processes as possible should be regulated by means of the market. The free play of market forces would result in the greatest possible benefit. *Laissez faire, laissez passer* was the legend emblazoned on their banner. This highly optimistic attitude to the market was accompanied by an even greater measure of mistrust of the state. This school of thought dictated that the tasks allocated to the state should be kept to a minimum, a set-up sometimes characterized as a 'nightwatchman state'.

When pondering the relationship between market and the state it was typical of the laissez-faire school to see these two fundamental institutions as two separate and mutually exclusive giants. Society could trust in the market or put its faith in the state but not do both at the same time. The school paid no heed to the historical and systematic interrelatedness of the two institutions. Another laissez-faire tendency was to present the market as a natural environment. The interaction between individuals in the market was not supposed to be mediated by institutions. Of course this perspective only served to reinforce the notion that the market and the state were at opposite poles to one another. After all, if the market is a natural domain then it hardly needs a state to mediate its existence. A last feature of laissez-faire thinking is more or less implicit in the above. The free market society was put forward as the best possible order. On the one hand this evaluation was based on the deontological notion that the free market formed a reflection of the natural and just order. On the other hand it was founded on the consequentialist conviction that from a practical point of view there was no better order possible. Any attempts by the state to come up with a better situation would be to no avail or would only make the situation worse (see also Hirschman, 1992 or Searle, 1998).

A grisly mental model

I only wish to embark on a brief evaluation of the laissez-faire mental model here. From today's normative perspective, it is easy to dismiss laissez faire as a grisly school of thought which lent legitimacy to the 'voluntary slavery' that characterized the labour market of the day, with its child labour and 16-hour working days. Although partially justified, this normative evaluation also removes laissez-faire thinking from its historical context. Cobden's laissez-faire thinking for example was also strongly inspired by the pursuit of world peace. Cobden saw the British government to a large extent as an instrument in the hands of a bellicose aristocracy. He therefore thought that minimizing political relations would considerably increase the chance of building a peaceful international community.

However, if we subject laissez-faire thinking to analytical scrutiny, we are forced to admit that it lacked structure and substance. The arguments put forward against all forms of state intervention were often far from convincing or in any case not well thought out. The same was true of its view of the market. The conviction that the market would serve the public good by means of the invisible hand was never supported by a serious line of argument, nor were the possibilities and limitations of the market ever thoroughly mapped out. What is more, the notion of the market as a just, natural order often remained implicit and was therefore weakly supported.

The rise of a new mental model of social organization

The laissez-faire mental model reached its peak between 1840 and 1870 (see Polanyi, 1944; Searle, 1998). Even as late as the turn of last century, the star of Cobden and Mrs Marcet was still a fixture in the firmament. Nonetheless, by the time 1900 rolled around, the night in which laissez faire had ruled was not to last for much longer. By then a tumultuous dawn chorus made up of academics of all persuasions had assembled to sing the praises of the new desired societal order. This choir included an important group of new economic authors, among them Knut Wicksell, John Maynard Keynes and Arthur Pigou. The ideas of these new economists were very much in harmony. All of them maintained that *in the real world* laissez faire would not automatically lead to a *social* optimum. Under circumstances that were less than ideal, they felt that market forces would have socially unacceptable consequences if left unfettered. And what is more, they even felt that laissez faire would fall short of the *economic* optimum in the real world. Despite these similarities, Wicksell, Keynes and Pigou all had their own specific points of emphasis. Wicksell focused on the social and economic necessity of social politics. Meanwhile, Pigou and Keynes primarily underlined the market's lack of success when it came to realizing its *economic* pretensions.

Academics like Wicksell, Keynes and Pigou brought about a shift in the way economists thought about the mental model of social organization. In order to illustrate this new wave of thinking, I will go on to discuss Pigou's theory in greater depth. In the context of this study, Pigou is the logical choice. He represents an important link in the neoclassical tradition. What is more, he is seen as the patriarch of the neoclassical environmental economists, given that he was one of the first writers to systematically investigate the economic aspects of environmental problems.

Arthur Pigou versus laissez faire

Many regard Pigou as the first great environmental economist. However, the environment was not Pigou's primary concern. Environmental problems come into play as part of his critique on laissez-faire thinking, which takes up a substantial part of his major work, *The Economics of Welfare*. He felt that the laissez-faire mental model of social organization was in dire need of revision. It was becoming untenable simply to maintain that '*the free play of self-interest will automatically lead to ... a larger output and therefore, more economic welfare than could be attained by any (other) arrangement ...*' (Pigou, 1920: 127).

To start with, Pigou pointed out that *free play* is best suited to aims that *'can be brought directly or indirectly into relation with a money measure'* (Pigou, 1920: 31). As he saw it, there were a great many social objectives which did not fit into this category. Although Pigou did not specify the exact nature of what he consistently referred to as these 'higher' cultural, normative and political objectives, his desire to include a range of environmental problems in this group seems clear. This impression is particularly strong when he discusses the problems relating to the environment as a long-term production factor, mentioning for example the problems of *'exhaustible natural resources'* for future generations (Pigou, 1920: 29).

The critique of the applicability of the market for achieving non-economic objectives only forms an introductory manoeuvre in *The Economics of Welfare*. Pigou's main concern was to show that, even in terms of *economic goals*, laissez faire could not always be defended as providing the most appropriate path. Like Keynes, Pigou felt that even this limited claim did not really hold water.

In support of his position, Pigou first set about gathering the necessary empirical evidence. He systematically charted all those areas of economic life in which the free play of conflicting self-interests failed to result in the best possible situation. He focused extensively on agriculture, arguing that farming in Britain at the beginning of the 20th century would have benefited from structural long-term investments, for example in buildings or the infrastructure of the land. However, these failed to materialize. According to Pigou there is a good explanation for this: the situation of the tenant farmers was fraught with uncertainties. Most worked on the basis of short-term contracts which did not oblige the landowners to compensate them for any long-term investments they made within the term of their contract.

A second area to which Pigou devoted a significant amount of attention was technological innovation. Pigou viewed innovation as a major driving force behind the expansion of economic production. A laissez-faire market, he argued, would only innovate slowly due to the high level of investment involved. On such a free market, no one would be particularly eager to lead the way in making such investments since the risk that competitors would jump on the bandwagon and take advantage of the resulting innovations would be too great.

Having established farming and technological innovation as two examples of areas where an economy based on free play structurally fails to take advantage of potential economic benefits, Pigou then took this argument even further. He pointed out that a free-play economy can easily give rise to situations where actors are not compensated for losses suffered. It is in this context that Pigou brought in environmental problems. He described clean air, clean water, attractive natural surroundings and an appealing landscape as economic goods, on the one hand because the environment is a long-term production factor and on the other hand because it is an economic consumer good. He therefore conceptualized any negative influence on these goods as a cost item. The value of some of these goods can only be indirectly linked to the money rod, yet Pigou did not regard this as a reason for abandoning this conceptualization. After all, damage to the environment can be very tangible and can sometimes be expressed in financial terms without having to resort to all kinds of artificial constructions.

'The same thing is true of ... smoke from factory chimneys: for this smoke in large towns inflicts a heavy uncharged loss on the community, in injury to building and vegetables, expenses for washing clothes and cleaning rooms, expenses for the provision of artificial light, and so in many ways.' (Pigou, 1920: 184)

Pigou did not limit himself to an empirical inventory. He also gave a systematic explanation for the occurrence of situations in which a society with a free-play market failed to pluck the fruits of people's egoism. According to Pigou, the free play of self-interest could only lead to an economic optimum if the actions an individual has to take to benefit society are exactly the same as those which best suit the individual himself. After all, the market mechanism never serves the social optimum directly. The social optimum 'hitches a ride' on self-interest. However, the mechanism falters when an individual's profit and loss account and that of society start to diverge.

To clarify his point, Pigou introduced a pair of concepts: 'marginal private net product' and 'marginal social net product', borrowed from Alfred Marshall, his illustrious predecessor at Cambridge. The marginal social net product is the *total* value of a marginal decrease or increase in economic activity expressed in financial terms, *'no matter to whom any part of this product may accrue'* (Pigou, 1920: 134). The marginal private net product is *that part* of the total financially expressible value of a marginal change in economic activity which flows directly to *'the person who is responsible for the change'*.

If free play is to have no disadvantageous effect on the social net product, then the marginal private net product of an action may not be greater than the social net product. However, if free play is to result in the social optimum, then it has to fulfil an even stronger claim: the most favourable action for the individual must coincide with the action which is the most favourable for the collective. Actual economic activities do not meet the first demand in many instances, and they meet the second even less often. In all these cases there is a discrepancy between the social and the private net product: an actor is encouraged to take action or refrain from doing so on the basis of his own individual economic benefit even though this is far from being the most beneficial course of action for society as a whole, since it leads to a net loss. Air pollution is a good example of this phenomenon. It constitutes a loss for society but generates profit for the polluter by allowing him to produce more cheaply. Excessive deforestation is another example. Deforestation on this scale leads to a loss of forest and ultimately to erosion problems. Yet a concession holder makes a tidy profit from the sale of the wood and remains unaffected by the long-term damage.

Pigou versus naturalism

Although Pigou's concept pair of 'private net product' and 'social net product' are no longer in common use, it is obvious that Pigou's theory is still very much alive today. The hand of the master can clearly be discerned in the modern theory of positive and negative external effects. It is partly for this reason that Pigou is revered by environmental economists. He was one of the first to conceptualize environmental problems in economic terms and to systematically show the impossibility of the laissez-faire market serving as the guardian of the environment. This being the case, it is all the more striking that another aspect of Pigou's criticism of laissez faire has completely fallen by the wayside. I refer to Pigou's critique on the notion that the market is a natural sphere. According to Pigou, advocates of laissez faire never tire of suggesting that free play would 'automatically and naturally' create the social optimum. Pigou regarded this naturalistic undertone as inappropriate. He was quick to emphasize that the market is an institutional construction, and that the favourable working of the market is nothing more or less than the product thereof. It was therefore with wholehearted approval that he quoted the economist Cannan:

> '*It has been said by a recent author that "the working of self-interest is generally beneficial, not because of some natural coincidence between self-interest of each and the good of all, but because human institutions are arranged so as to compel self-interest to work in directions in which it will be beneficent*"' (Pigou, 1920: 128).

Even Adam Smith's record in this respect was not entirely unblemished. Pigou felt that Smith failed to adequately realize the extent to which '*the system of natural liberty*' had to be adjusted and safeguarded by laws and regulations before it could ensure that the factors of production were put to productive use. As we will see in the next section, this part of Pigou's criticism not only applies to his forerunners but also to his heirs.

Pigou and the competition between the market concepts

Earlier we identified three conceptualizations of the market: the empirical-historical definition, the functional definition and the model-based definition. In the first definition, the market is described in terms of a systematic overview of the characteristics which have proved to be significant in its development. The second, functional account, establishes the market as a sphere where, by definition, public goals and private goals harmonize. In the third, model-based definition an ideal or typical market model forms the starting point and the market is defined as a sphere which meets the parameters of this model (full competition, full information, and so on).

It is useful to apply the distinction between these market descriptions to the analysis of Pigou's critique on laissez-faire thinking. The most valuable and analytically powerful aspect of Pigou's thinking about the difference between social and private net product was that it broke the automatic link which laissez-faire thinking had forged between the empirical-historical and the functional market concepts. Pigou disagreed strongly with the notion that a market formed in accordance with the empirical-historical definition would lead to a situation in which public and private interest come together. The two descriptions do not always refer to the same phenomenon but are sometimes in competition with one another. Pigou saw the evidence with his own eyes: the empirically-historically fashioned market of his time had not resulted in an economic optimum but in environmental problems, postponement of necessary investments and so on.

The insight that these market concepts had to be separated from one another obviously presented Pigou with something of a dilemma. The incompatibility of the market concepts implied that Pigou had to choose between them. He had to decide which to use as his primary concept. Reconstructing Pigou's theory, we can state that in principle he had three options in this respect.

First of all, he could have taken the empirical-historical market concept as the starting point for his theory. The advantage of this choice was that it would have given him a clearly defined market concept. However, it would also have meant abandoning the claim that the market leads to the creation of social harmony. He would then have been forced to concede that the market is only a 'limited use institution', of possible economic value but by no means beyond all criticism.

Pigou's second option was to pursue the further development of the model-based market concept. This option had two major advantages. It too offered clarity in terms of concept definition while also retaining the claim that the market leads to an economic optimum. The disadvantage inherent in this option, however, was that it complicated the relationship to reality, creating the need for all manner of auxiliary theories to translate the findings and conclusions from the model into reality.

This left the functional market concept as the third option. The major advantage of this choice was that it encompassed the claim that the market is a sphere in which private and public interests coincide. But it too had its drawbacks. Opting for the functional market concept meant losing sight of the concrete design of the market. What were the empirical characteristics of a sphere where public and private interests coincide? This could vary considerably depending on time and place. In some situations, for example when few external effects are in play, the market would probably have a legislative regime that allocates limited liability to actors, while in other situations an alternative legislative regime could well be desirable or indeed necessary.

For me, the special thing about Pigou is that he opted for this last alternative. He regarded the market first and foremost as the sphere in which public and private goals should come together. In his eyes, the challenge to politicians and academics was to ensure that the institutions were formed in such a way that this ideal would become reality.

A Pigovian revolution

By pledging allegiance to the functional market in the face of these competing market concepts, Pigou brought about a shift in the neoclassical mental model of social organization. For although Walras, Pareto and other early neoclassical authors did not really defend laissez-faire thinking as such, their mental model of social organization did not criticize it explicitly. Pigou, on the other hand, is miles away from laissez-faire thinking. In his view, the market is no longer a natural sphere. Instead he conceptualized it as an institutional construction, a sphere which is not simply and spontaneously present whenever people interact, but which first has to be created. Another, related point is that Pigou's arguments also created a shift in thinking about the *relationship* between market and the state. According to Pigou's way of thinking, market and the state were no longer opposed to one another and no longer mutually exclusive.

For Pigou left no doubts as to who was responsible for the adequate design of the market as an institutional construction. He placed this task firmly in the hands of the state. It was up to this fundamental institution to ensure that the market became the sphere in which public and private interest could be brought into harmony:

> '*It is the clear duty of government, which is the trustee for unborn generations as well as for its present citizens, to watch over, and, if need be, by legislative enactment, to defend, the exhaustible natural resources of the country from rash and reckless spoliation*' (Pigou, 1920: 30-31) ...

> '*It is, therefore, necessary that an authority of wider reach should intervene and should tackle the collective problems of beauty, of air and of light, as those other collective problems of gas and water have been tackled.*' (idem, 195).

In Pigou's thinking, the state was accorded a much more demanding task and a far greater responsibility than could ever have been conceived along the lines of laissez-faire thinking. Pigou placed the state at the heart of his theory. It was up to the state to constitute the market and so it was also largely up to this institution to ensure that the market brought the self-interest of its actors into harmony with the common good. The danger that this task could stretch the capacities of the state to breaking point clearly did not occur to Pigou:

> '*It is, however, possible for the state, if it so chooses, to remove the divergence in any field by "extraordinary encouragements or extraordinary restraints" upon investments in that field.*' (idem, 192) .

> '*This adjustment of institutions to the end of directing self-interest into beneficial channels has been carried out in considerable detail*' (Pigou, 1920: 129).

Pigou's Achilles heel

How should we evaluate this new Pigovian mental model of social organization? In doing so, I think it best to return to Coase's critique on the neoclassical tradition. Coase aimed his criticisms explicitly at Pigou. He felt that Pigou took too little account of the limitations of the state.

When stated in such general terms, this criticism of Pigou does not hold up. In *The Economics of Welfare*, Pigou does in fact demonstrate his awareness of the failure of the existing state on several occasions. However, if we take Coase's critique and formulate it more precisely, it leads us straight to the flaw in Pigou's thinking. Pigou fails to see the limitations to the *potential* of the state. Apparently he could conceive of a state which would be capable of avoiding all the mistakes of which the state of his day was guilty. And what is far more damning within the context of this study of the mental models of social organization, is that Pigou's mental model actually depends to a very large extent on such a superiorly functioning state. All things considered, and despite the keenness of his analysis, it is hard to see Pigou's renewal of the neoclassical mental model as much of an improvement on laissez-faire thinking. Ultimately, Pigou exchanges the notion of a natural, perfect market for the notion of a perfectly functioning state. Ultimately it is an unsatisfactory exchange, especially when viewed from our modern-day perspective and our all too painful awareness of the limitations of the state.

2.6 THE MENTAL MODEL OF PRESENT-DAY NEOCLASSICAL ENVIRONMENTAL THEORY

Two economic mental models of social organization have been identified and analysed so far: the 19th century laissez-faire mental model and Pigou's neoclassical mental model which dates from the beginning of last century. Analysis of these two mental models reveals that neither holds up particularly well. Laissez faire turns the market into a miracle mechanism while for Pigou the state functions as a universal panacea.

In this section we turn our attention to the mental model of social organization adhered to by present-day neoclassical environmental economists. As we have seen, this group is eager to present its approach as a new vision on how to control the market. They want market-based regulation to take over from the bumbling state and its misguided attempts to get a grip on society using command and control tactics. It is therefore interesting and useful to probe the mental model of social organization which lies at the heart of the ideas surrounding market-based regulation.

The structure of this section is straightforward. It begins with a brief review of the mental model of social organization embraced by the neoclassical environmental economists. It then addresses the question of whether present-day neoclassical environmental economists have succeeded in transcending laissez-faire thinking and concludes with an examination of whether or not they have travelled beyond the ideas of Pigou.

The market as a limited use institution.

Modern-day neoclassical economists are remarkably reserved about the value of the market and the possibilities it offers. They steer well clear of the notion that all societal processes should be coordinated via the market. Even in the 1980s, the decade in which liberal-democratic worship of the market reached new heights, this radical notion elicited hardly no support among neoclassical economists. According to neoclassical thinking, the market is first and foremost an adequate mechanism with which to achieve society's economic goals (Samuelson and Nordhaus, 1985: 4). In addition to this, present-day neoclassicists are quick to stamp out any misconceptions about the market being free of restrictions and disadvantages, pointing out that in practical terms it has its fair share of imperfections.

Three areas often mentioned to demonstrate the *restrictions* (see Section 2.4) of the market are the defence of the nation, stability (particularly in the social-economic sphere) and the constitution of the law (see for example Samuelson and Nordhaus, 1986: 47-51). Where market *disadvantages* are concerned, many a neoclassical economist refers to the distribution of income (social justice). Most would recognize that even from a purely economic point of view it is not a good idea to simply give the market free rein. For one thing, such an approach would only worsen the periodic recessions to which the economy is subject (Eijgelshoven et al., 1993: 11-13). The insight that real markets suffer from *imperfections* because it is utterly impossible for them to meet the stringent conditions that apply to a model market is also fully accepted by present-day neoclassical economists. For proof we need look no further than the neoclassical environmental economists' theory of external effects.

All these qualifications do not of course take away from the fact that neoclassical economists still see the market as a powerful fundamental institution which acquits itself of its economic tasks efficiently and well. Just like Pigou and the protagonists of laissez faire, they therefore feel that society should make use of the market in order to achieve its goals as much as possible. At the same time, it is clear that modern neoclassicists also reserve an important place for the state in their thinking. For them, the market is a 'limited use institution' which has to operate in tandem with the state. However, they give no systematic account of the tasks that the state should fulfil. Instead, the state is repeatedly nudged to the fore in cases when the market is found wanting due to its restrictions, disadvantages or imperfections.

Beyond laissez faire?

The contours of present-day neoclassical mental model have been sketched above. Today's neoclassicists interpret the market as an adequate mechanism for the fulfilment of economic functions. Since the market inevitably suffers from disadvantages, restrictions and imperfections, a society that relies wholly on a market would soon become unbalanced. In order to counter the disadvantages, restrictions and imperfections of the market, present-day neoclassicists look to the state. In their view, it is up to this fundamental institution to compensate for all of the market's limitations. The position of the state within current neoclassical thinking can therefore be justifiably compared to that of a troubleshooter. Every time the market is hampered by its shortcomings, the state leaps into action.

The answer to the question of whether the neoclassical environmental economists transcend laissez-faire thinking would therefore appear to be a resounding 'yes'. They appear to have taken the lessons of Pigou, Wicksell, Keynes and their ilk to heart. In the thinking of today's neoclassical environmental economists, the market has been devalued from a universal mechanism for dealing with all or almost all social tasks to an adequate mechanism, yet one of limited use, which suffers from inevitable disadvantages and whose effectiveness is subject to a whole range of conditions.

Yet perhaps we should not be too eager to give today's neoclassicists our blessing. Their thinking is after all typified by an economistic approach (see Section 2.1) that some would term 'notorious'. This economism means that neoclassicists are actually much more pretentious and optimistic about the power and the value of the market than it would first appear when they state that the market is only a mechanism for maximizing economic well-being. For them, economic well-being is more or less identical to 'the good life'. In this light, present-day neoclassical environmental economists are still a long way from getting beyond laissez-faire thinking. At the heart of economism, the idea that the market is a miraculous mechanism which realizes maximum well-being behind the backs of the people still lingers.

Several other beliefs held by today's neoclassical environmental economists can only be properly understood when we assume that the embers of laissez-faire thinking smoulder on in their thoughts. One example is the tendency to see the functional, the model-based and the empirical-historical definitions of the market as extensions of one another. We see this happen, for example, when the advocates of market-based regulation governed by private law propose the creation of private property as a problem-solving strategy. Here two concepts of the market are inadvertently compounded. On the one hand, the market is seen as a place where public and private interests necessarily coincide. On the other hand, the question of what structure this market should take is answered with recourse to the model-based conception of the market.

Viewed from another angle, one can say that neoclassical environmental economists today pay too little heed to 19th century developments in thinking about the free market. In the second half of the 19th century, a remarkable parting of the ways took place among theorists grappling with the legitimation of the free market. A split occurred between the utilitarian defence of the free market and legitimation on the basis of the natural rights (Searle, 1998). Until around 1850 these two views coexisted peacefully: the idea of a free market was justified by pointing out the benefit that this fundamental institution had for society as a whole and in the same breath by referring to the importance of inalienable rights like freedom and ownership.

As the 19th century progressed, however, various theorists began to realize that these two justifications could not always go hand in hand. In the face of poverty, external effects and so on it became impossible to maintain that the empirical-historical system of exchange, limited responsibility, ownership etc. brought maximum benefit to all. Slowly but surely this insight gave rise to two different justifications of the free market, which continue to exist separately from one another to this very day (Koslowski, 1986; Sen, 1993).

The first justification appeals to the empirical-historical market concept and is in essence an argument for the inalienable nature of rights such as liberty and ownership. The other justification draws on the functional market concept and justifies the market by emphasizing its benefit to society. It could be argued that the protagonists of laissez faire found themselves in the thick of this process and may therefore have conflated the various market definitions. Pigou was aware of the need to avoid this mistake but unfortunately his insight appears to have been lost on the present-day neoclassical environmental economists. The various descriptions of the market have become entangled once more, an error compounded by the development of the model-based market concept.

There is yet another important component to the theory of the present-day neoclassical environmental economists in which the influence of the laissez-faire thinkers shines through. This concerns their naturalism and - another closely related aspect - the way in which they place market-based regulation and the state in opposing camps. Once again it could be argued that, in any case after Pigou, this simplistic antithesis should have disappeared. The relationship between the market and the state is far more complex than this simple oppositional model allows. Of course there are ways in which the market and the state can be viewed in opposition to one another: the manufacture of a product like shoes can either be left to the market or ordained by the state. However, when we start talking about market-based regulation and the functioning of a market in a modern society then this antithesis no longer holds. In that case, market-based regulation implies a state which draws up rules and creates conditions whereby the market can function.

There is too little scope for this symbiotic or perhaps parasitic relationship between market and the state in the thinking of present-day neoclassical environmental economists and to my mind this does damage to their theory of market-based regulation. For if the market depends on the state in order to function, then this automatically undermines the idea that market-based regulation can serve as an alternative to control by the state. In this sense the theory of market-based regulation promises far more than it can deliver. The theory contains a hidden assumption. As soon as this assumption is unearthed, it confronts us with the question of whether market-based regulation actually lends a helping hand to our overburdened state or whether it simply presents the state with the same management problems all over again by means of another route.

Beyond Pigou

Do today's neoclassicists succeed in getting beyond Pigou? Pigou made a theoretical miscalculation by proposing that the problems associated with the miraculous mechanism of the market can be solved by presenting the state as a miraculous mechanism. As has already been implied above, today's neoclassical environmental economists have failed to correct this slip. A telling sign of this failure can be found in the fact that the present-day neoclassicists do not complement their systematic analyses of the restrictions and the disadvantages of the market with a comparable analysis of the state. The state still features in neoclassical thinking as a troubleshooter whose potential is not to be doubted. A neat illustration of this unshakeable confidence in the potential of the state can be found in *The Economics of Environmental Policy*, where Freeman III, Kneese and Haveman identify a number of '*urgently needed steps*':

'1. *Wherever possible, taxes or charges should be imposed on the use of common property resources for residuals discharge.*
2. *Institutions should be developed to manage river basins and airsheds* [Kneese's case studies] *on an integrated basis. These institutions should have genuine authority and at the same time be accessible and accountable to the people.*
3. *Adequate government support for research on all aspects of the environmental quality management system should be provided.*

> *4. The political process should be improved so as to increase its visibility, accessibility, accountability, and representativeness.'* (Freeman, Kneese and Haveman, 1973: 173).

This is a wonderful example of a 'to do' list for the modern liberal democratic state. However, absolutely no mention is made of *how* the state is supposed to go about carrying out these tasks. Freeman III, Kneese and Haveman apparently see no real obstacles in that quarter. It is exactly this blind optimism that makes their list more like an incantation than a relevant contribution to the debate about managing environmental problems.

But aren't economists also quite willing to level serious criticism at the way in which the state *actually* functions? This is indeed the case, but to my mind this does not necessarily represent a dilution of their trust in the *potential* of this fundamental institution. In a sense, such criticism could even be seen as a confirmation of my position: the criticism aimed at actual performance appears to be inspired by the belief that it could all be done so much better.

This becomes most apparent when we consider the way in which economists explain the current failures in the state. Structural causes typically remain tucked out of sight in economic explanations for state action. Instead lobbyists, short-sighted politicians and/or power-hungry bureaucrats hog the spotlight in neoclassical accounts of state failure. The institutional make-up of the liberal democracy, and in particular the relationship between the market and the state, remain at a safe distance from the firing line. All the usual explanations come down to the notion that individuals working within the state have put their private interests before the public interest.

> *'First we tend to think, as citizens, that the duty and purpose of governments is to act in our interests as a community rather than as individuals. This is why we have laws, police forces, a judiciary, public health regulations and so on; but this image of 'benign' governments can be false. At the one extreme, governments may be despotic and interested only in favouring the interests of some part of the community rather than the community as a whole. Even in democratic countries, governments may act as to please a particular pressure group rather than the community as a whole. This means that governments may well not act as to protect the environment, especially if they think that environmental protection will impose costs on members of powerful pressure groups. ...*
> *Third, government, in the form of politicians, may have good intentions and frame a good environmental law in principle. However, it has to be translated into practice and this involves using experts who are part of a government bureaucracy. The bureaucrats become very important and can easily influence the nature of the regulations in practice. Since bureaucrats are very often not elected officials and, unlike many workers, they tend not to be paid by results, they therefore have little explicit incentive to behave in the best interest of the community unless closely scrutinized by politicians - and that can be very difficult.* (Pearce and Turner, 1994: 80).

All in all, I think it is reasonable to conclude that although present-day neoclassical environmental economists are aware of the restrictions, imperfections and disadvantages of the market, there are still quite a few snags attached to their mental model of social organization. From a highly critical viewpoint, they can be said to combine the worst of laissez faire with the worst of Pigovian thinking. Today's neoclassical environmental economists, like the laissez-faire thinkers before them, mix up the various concepts of the market and set up an exaggeratedly one-dimensional opposition between the market and the state. In doing so they implicitly follow Pigou in transforming not the market but the state into some kind of miraculous mechanism.

A logical question to ask at this stage is whether I am not simply exaggerating matters. Do we really overestimate the state by entrusting it with the task of ensuring that all the right limiting conditions are in place for the market to function adequately? The answer to this question depends to a considerable extent on the concept of the market one has in mind. If we take the empirical-historical market concept as our basis then in principle the state is not really faced with insurmountable difficulties. In the course of a few centuries' experience, the state has learned how to constitute the market as a system is based on freedom, exchange, limited responsibility, competition and ownership.

However, the situation is very different when operating on the basis of a functional market concept, as today's neoclassical environmental economists do, at least in part. For operating on the basis of such a concept implies that the market as a mechanism is desirable and justified only insofar as it serves the public interest. In the functional market concept, it is the *bringing about* of social harmony between the individual and the public interest which is the market's *raison d'être*. Accordingly, the state not only has to guarantee freedom, competition and the like but it must also keep on tinkering with the institutional structure of the market until the sum of the inadvertent and intentional consequences of everyone's actions within the framework of their own interests maximizes the benefit to all.

This task of harmonization is not one to be taken on lightly and it is the enormity of this task that leads me to predict the overburdening of the state in a free market society which operates on the basis of this concept. We only have to think of the information needed to map out the intended and unintended, present and future consequences of everyone's actions, never mind analysing it to come up with the relevant incentives aimed at influencing the behaviour of individuals in just the right way. It is easy to see how the need for this data alone could take on staggering proportions. But it doesn't stop there. The state would also have to obtain information more rapidly than those who are active in the market and interpret it more effectively in order to pre-empt anticipatory changes in conduct.

In short, in a society where the functional interpretation of the market rules, the very value of the market as an institution becomes questionable. One of the main arguments favouring the free market over a communist system after all is that the free market requires little in the way of centralized information (Hayek, 1976). This justification of free-market superiority evaporates in a society where the state has an insatiable need for information in order to keep the market working properly.

As I see it, this argument becomes even more powerful within the context of the current debate about managing collective objectives like sustainability. There the advocates of market-based regulation are continually trying to utilize the market as a functional institution for non-traditional and even atypical economic goods like clean air and clean water.

2.7 LOOKING BACK AND LOOKING AHEAD

In this study, my first aim is to find out whether the indirect responsibility model still has a future. My second goal is to examine the form an alternative, direct responsibility model might take. These questions will be answered by way of a detour. In order to arrive at these answers, I am examining the ways in which three separate academic disciplines think about controlling the market, in the light of environmental problems. This chapter analysed the theory of the neoclassical environmental economists. What has this analysis achieved in its own right and what does it have to contribute to the main aims of this study?

Many people, both academics and non-academics, feel that the existing institutional order of modern liberal democracies is flawed. Consequently, there is a widely held belief that environmental problems are a structural component of these societies. The powerlessness of the state is often presented as the main structural defect of modern Western society. The state appears to lack the power to solve public problems and channel social processes in a positive direction.

The theory of the neoclassical environmental economists is in perfect keeping with this general feeling. Neoclassical environmental economists claim to have an alternative to traditional state control of the market. However, determining the exact nature of their claim is not that easy. On the one hand the neoclassical economists would appear only to be arguing for a new set of measures that the state can use to control the market. On the other hand they appear to want a whole new relationship between state and market or, to use the terminology of this study, they want a whole new mental model of social organization. This study bases its analysis of the neoclassical environmental economists' theory on this second claim.

At the heart of the new mental model designed by the neoclassical environmental economists is the notion that the market should take over responsibilities from the state. The market should be organized in such a way that actors are driven by their own self-interest to realize sustainable development, thereby ensuring that the pursuit of private interest will lead to public benefit.

This study firmly rejects the above model, for a number of reasons. First of all, analysis reveals that neoclassical environmental economists simultaneously employ three different conceptions of the market: the historical-empirical, the model-based and the functional concepts. This mix-up leads to inconsistencies. Secondly the neoclassical environmental economists have ultimately been unable to prove their claim that their new mental model unburdens the state. Their construction is based on the silent premise that it is up to the state to establish a market that works perfectly. Accordingly their approach rests on the idea of a state that works perfectly. A third, closely related objection is that neoclassical economists fail to pay enough attention to the *structural* limitations of state power in modern-day society.

Placed in the light of the general aims of this study, the analysis in this chapter has provided initial proof in support of the claim that the indirect responsibility model is in trouble. After all, the thinking of the neoclassical environmental economists is fully compatible with this mental model. Their thinking reflects the three cornerstones of the indirect responsibility model in all its glory. The neoclassical environmental economists think that actors in the market only have a limited responsibility for public issues and they feel that this is the way it should be. In addition they want the market to be controlled by means of limiting conditions and they think that the full responsibility for solving public questions ultimately lies with the state. What we need to find out is whether the problems with the neoclassical mental model are specific to this tradition or whether they can somehow be linked to the indirect responsibility model in general. The next chapter, with its focus on the administrative theorists, promises greater insight into this issue. Administrative theorists, and Dutch administrative theorists in particular, have been fighting a crusade against the indirect responsibility model for years. They regard this mental model as completely outmoded. Let's see if their arguments pass muster.

CHAPTER 3

ADMINISTRATIVE THEORY:

From overload to taking responsibility

3.1 THE ADMINISTRATIVE PROGRAMME

Let us consult Herman Daly on the subject of controlling the market. Herman Daly is an internationally renowned advocate of sustainability, who also maintains that the market stokes up environmental problems (Daly and Cobb, 1989). An important strand in Daly's thinking is that the market should once again be brought under the control of politics and society. From a situation in which social life is encompassed by the economy, we must return to a situation in which the economy is encased in the entire interplay of social relations (Daly and Cobb, 1989:8). In contrast with so many other environmental philosophers, Daly makes it clear that he does not wish to account for this endeavour in romantic terms. A market cannot be led by civil society alone. His suggestion is that society should try to put into effect Adam Smith's vision on the subject. According to Daly, this means that the market is reined in by politics (i.e. the state). Daly takes careful account of liberal-democratic sensitivities in presenting his case: he does not advocate process management (see Section 2.2). 'Returning to the market as it was intended', involves working towards a society in which the state sets the limiting conditions for the market process.

It is interesting to compare this approach, which is so typical of the indirect responsibility model, with the vision of Dutch administrative theorist Jan Kooiman. Kooiman's vision exemplifies that of many of his colleagues from the Netherlands and beyond. One of the first things to strike the reader about Kooiman's work is the lack of hardline terminology about the state using laws and regulations to control the market. Instead, Kooiman places the emphasis on cooperation between the state and market actors, and he insists on how important it is for market actors to show a sense of their own responsibility. Of course, this does not mean that Kooiman completely rejects the operating procedures of the indirect responsibility model, such as the drawing up of regulations. However, this aspect of the work of the state does not feature prominently in his view of things (Kooiman, 1993a: 35-50 and Kooiman and Van Vliet, 1993: 58-72).

Daly and Kooiman both regard sustainability as an important social issue. While their normative insights on this matter do not differ greatly, they do, however, part company in their thinking on how the market should be controlled. Daly believes that the state harbours significant management potential, while Kooiman believes there are limits to the possibilities for controlling the market, especially when society wants to resort, perhaps even exclusively, to practice consonant with the indirect responsibility model.

This contrast between Daly and Kooiman can shed considerable light on the developments within the academic discipline of public administration, particularly in the Netherlands over the last forty years. In a nutshell, administrative theorists have set themselves the task of placing government control (including management of the market) back on the academic and political agenda (see for example Bovens et al., 1995; Den Hoed, 1979; Nelissen, 1992 and Van Vliet, 1992). Their most important message in this connection is that, in modern societies like the Netherlands, the mental model being used as a template for thought and action for government control no longer fits the context of the modern age. As a result, the second half of the 20th century has seen a chasm develop between political-administrative resolutions and political-administrative reality. In order to change this situation, the administrative theorists recommend that a new approach be taken. A new way of *dealing* with public issues has to be found. This implies that a new way of *thinking* about controlling the market also has to be developed.

In the meantime, academics have given many different names to the mental model (and the empirical situation) that they criticize. The terms 'command and control' and 'hierarchical-bureaucratic management' have already been mentioned; other commentators speak of 'rationalism' or a 'top-down approach' (Blansch, 1995: 22). Terms like 'constitutionalism' (Drucker, 1974), 'authoritarian-instrumental control model' (Pot, van der, et al. 1995: 538) and 'Copernican management model' (Hafkamp and Molenkamp, 1990: 247) are also in vogue. We will stick to our concept of the 'indirect responsibility model'.

Positioning this chapter in relation to the general aims of this study, we can see that it clearly relates to the first key question concerning the validity of the arguments against the indirect responsibility model. Present-day administrative theory in the Netherlands and beyond can be interpreted as one big critique on this mental model. There is therefore a good chance that this chapter could afford us a clear insight into the problems which might be inherent in it. However, the study of present-day administrative theory on controlling the market is also relevant to the second research objective. Administrative theorists also try to formulate alternatives to the indirect responsibility model. One of these attempts, co-management, will be examined in some detail.

Nevertheless, this chapter is more than just a presentation of current thought in public administration theory. To begin with, the critical nature of this presentation sets it apart from its more traditional counterparts within the world of administrative theory. I will argue that the administrative critique of the indirect responsibility model shows evidence of a number of significant flaws, for example with regard to its historical focus. Another distinctive aspect of this chapter is its consistent critique of the indirect responsibility model as a mental model. In a great many administrative studies it is never really clear whether the indirect responsibility model is being criticized as a way of thinking or as a practice. Given that, in a country like the Netherlands, the indirect responsibility model has been a far greater presence as a way of thinking than as a practice, this lack of a clear-cut focus leads to confusion (see Section 3.5). Yet another distinguishing feature of this chapter is closely related to this point. In this chapter the indirect responsibility model is consistently approached as a model pertaining to the *relationship* between market, state and civil society. This approach allows us to analyse the structural causes behind the problems associated with indirect responsibility model and avoid an analysis of these problems simply in terms of the supposed stupidity, bureaucracy, inflexibility etc. of state action. The latter is a pitfall that administrative theory does not always manage to sidestep. Some administrative theorists let their analysis degenerate into taking pot shots at the state.

The most important distinguishing feature of this chapter, however, is the fact that the indirect responsibility model is addressed in relation to *normative* thinking about liberal democracy. This normative link to liberal democracy usually fails to come to life sufficiently in administrative analyses. As a result, these analyses fail to demonstrate clearly enough how well the indirect responsibility model and liberal democracy go together and avoid the question of whether tinkering with the indirect responsibility model is in fact compatible with the values of a liberal democracy. In short, discussing the indirect responsibility model in its normative context makes it possible to articulate the strengths of this mental model and allows us to avoid the pitfalls of proposing solutions which fail to take sufficient account of liberal-democratic sensitivities.

One last question remains unanswered in this section: who exactly are the administrative theorists being spoken about here? They are not as easy to identify as the neoclassical environmental economists from the previous chapter, who form a group within a reasonably homogenous school of thought. When we talk about administrative theorists in this chapter, we are primarily referring to a loosely related group of administrative theorists who are concerned with the issue of controlling the market in relation to the issue of sustainability. The main figures in this respect are Dutch theorists like Kooiman, Hafkamp, Glasbergen and Van Vliet. This group is not restricted to the Netherlands, however. International authors who meet these criteria include Renate Mayntz (1978), Sven Jentoft (1989, 1995, 1998, 2001) and Bardach and Kagan (1982).

The administrative theorists who address this topic of sustainability and the market, however, are not a group unto themselves. In their work they repeatedly refer to other administrative authors. In order to strengthen their position, therefore, I have also enlisted support from authors belonging to this 'outer circle'. They include the Dutchman Mark Bovens and the English author Christopher Hood. We should also not overlook the fact that administrative theory maintains intensive contacts with other academic fields, such as law and sociology. Accordingly I have also referred to authors from these adjoining disciplines in my efforts to state the case against the indirect responsibility model as strongly as possible. Examples here include the likes of the American philosopher Albert Hirschman, the American legal sociologist Peter Yeager and the German legal philosopher Ingeborg Maus.

3.2 THE INDIRECT RESPONSIBILITY MODEL

'(It) depends ... on the limiting conditions *(of the market) whether the pursuit of self-interest will lead to a collective catastrophe or contribute to the common good. It is therefore the task of the government's economic policy to organize the limiting conditions in such a way that this second possibility becomes reality. A government that does not establish limiting conditions in this way makes itself culpable - for in an economy based on competition it is only possible in exceptional circumstances for an individual company to act in a way that promotes the common good if this entails an economic disadvantage for the company itself.'*

(Hösle, 1996: 88; passage emphasized in original text)

'To impose rules as to the way a business should be run is going too far. The way a business is run and the character of a business operation is primarily the responsibility of the businessman. ...(The Minister) is only able to and only willing to establish limiting conditions.'

The Dutch Minister of Agriculture, Nature Management and Fisheries
Eerste Kamer der Staten Generaal, 1991-1992: 7.

In the next two sections, an attempt will be made to reconstruct the indirect responsibility model. A methodical problem one is likely to encounter when embarking on such a task is that mental models of social organization are by definition vague and open to multiple interpretations. Expressing such a mental model in explicit terms is therefore a precarious undertaking. Many people think about social organization along the lines of the indirect responsibility model. The above quotations are examples of this. But, almost by definition, no one does so in an explicit and self-conscious way. The only ones who do speak explicitly about the indirect responsibility model are its critics, but they oftentimes set it up as a man of straw, with the express intention of tearing it down. What makes this undertaking especially tricky is that I have set myself the aim of *enhancing* the indirect responsibility model. By forging the link with liberal democracy, I wish to present the indirect responsibility model in such a way that it has the opportunity to defend itself against its critics.

As I see it, it is best to take an ideal-typical approach to this reconstruction. This means setting out all the essential elements of this way of thinking as strongly and systematically as possible without claiming that all the interpreters of this tradition, past and present, adhere to this particular version. Or to go a step further, it can be described as a version which *'in this absolute, ideally pure form is probably encountered as little in reality as a physical reaction calculated on the assumption of an absolutely empty space'.* (Weber, 1921: 10).

A mental model of social organization consists of a description of the internal structure of the three fundamental institutions of society (state, market and civil society) and also describes the links between these fundamental institutions. This chapter is mainly concerned with describing the state and the links between the state and the other fundamental institutions. I will refrain from discussing the indirect responsibility models' vision of the market here. This has more or less been done in my account of the neoclassicists in the previous chapter, for, as I already mentioned, they turn out to be the ultimate representatives of the indirect responsibility model. I will also more or less bypass civil society in my account and with good reason. Under the indirect responsibility model, civil society hardly has any relevance in terms of controlling the market or social management in general. The indirect responsibility model's notion of social management focuses exclusively on the relationship between the market and the state. The indirect responsibility model only allocates a role to civil society when it comes to organizing democracy.

In this section I will first give a short and schematic overview of the indirect responsibility model. In this section and the next, I will then proceed to develop the model in greater detail with regard to a number of important points. In doing so it is my aim to reveal the close relation between liberal democracy and the indirect responsibility model. For this reason my argument will sometimes take a rather general approach to the exploration of liberal-democratic thinking as such. For the sake of clarity I will restate the specific aspects of the indirect responsibility model at the end of each section.

The most important characteristics of the indirect responsibility model are the following:
* *The market is regarded as a sphere in which actors bear a limited responsibility for public issues.*
   In the indirect responsibility model, the market is seen as a domain in which people pursue their own self-interest and are expected to do so. This serves to benefit the logic and therefore the efficiency of the market process. This orientation towards self-interest also contributes towards realization of consumer sovereignty and the ideal of democracy (see below).
* *The state has an almost exclusive responsibility for public issues.*
   The indirect responsibility model assumes a strict division of labour between market and state. This division not only benefits the rationality of the market but also strengthens democracy. If all public issues are regulated by the state, then this gives expression to the ideal that all citizens should have the right to contribute to the decisions made with regard to all public issues, either directly or indirectly. After all the state is under the control of the citizens.
* *The state controls the market by means of limiting conditions.*
   This form of government means that the state only issues general rules and in doing so marks out the domain within which economic actors enjoy freedom of action. This form of control not only fits in well with the rationality of the market but also with the ideal of a free society in which the state does not control the lives of free citizens. This form of government also helps realize the ideal of a pluralist society.
* *The state should be a fettered giant.*
   According to liberal-democratic thinking the state should be powerful enough to carry out its tasks. At the same time, its power needs to be limited in order to combat the grave dangers associated with a state that has too much power. The indirect responsibility model solves this paradox by such means as the ideal of the *Rechtsstaat* and the distinction between private and public.

*   *The distinction between public* issues *and private* issues *comes together with the*
    distinction between the public domain *and the private* domain.
    The distinction between public and private is fundamental to liberal-democratic
    thinking. The distinction is made up of several layers, however. It is possible to
    distinguish between the public and private domain and between public and
    private issues. A private *issue* is a matter which only concerns the individual. A
    public issue is related to matters that concern all (see Section 1.3). The private
    domain is the domain in which actors are free to act as they see fit. It is the
    domain where there can be no power 'legitimately exercised by society over the
    individual' (see Mill, 1859: 59). The public domain is the domain in which
    society can legitimately exercise power. Since the state is the ultimate public
    actor in the indirect responsibility model, we can also say that the public domain
    is the domain where the state has the jurisdiction to act.
        In the indirect responsibility model both these dichotomies become one. The
    private domain is looked upon as the sphere where only private issues are at
    stake. Meanwhile public issues fall entirely within the public domain and should
    therefore be dealt with by the state. In relation to controlling the market, it is
    important to note that property is one of the rights that constitute the private
    sphere. The market therefore forms part of the private sphere to a great extent.
    Nevertheless, when faced with this idea, liberal democrats do not conclude that
    control over the market should be completely prohibited. This thought is
    mitigated to the idea that the state is only allowed to intervene in the market in
    specific ways. The state should restrict itself as much as possible to control by
    limiting conditions (see also Section 3.3).

*   *There are not many bridging or linking institutions between the fundamental*
    institutions.
    A bridging institution is an institution which links two fundamental institutions
    with each other and makes interaction possible. Parliament is an example of a
    bridging institution between state and civil society. The indirect responsibility
    model demonstrates great reluctance when it comes to the creation of bridging
    institutions. According to this mental model, such bridging institutions disrupt
    the rationality of the separate institutions. It follows that there ought not to be
    much interaction between the spheres when it comes to policy and legislation.
    This applies to all phases of the policy process from preparation to enforcement
    and implementation. Structures set up with the express purpose of facilitating
    such interaction are out of the question within this mental model.

*   *Democracy as representation in the state.*
    According to the indirect responsibility model, the citizens determine legislation
    in a true democracy, either directly or indirectly. To a considerable extent,
    'sovereignty of the people' is translated in terms of a representative system that
    puts the citizens in a position to influence and monitor elected governors and
    public representatives. In addition to this, certain forms of consultation are seen
    as inevitable in the modern context.

*   *Disregarding civil society when it comes to controlling the market.*
    The indirect responsibility model has very little regard for civil society when it
    comes to political and administrative matters. Electing administrators and public

representatives is the only role allocated to civil society. This limited role for civil society is the only facet of this mental model of social organization that is difficult of rhyme with liberal-democratic thinking. After all, self-government is an important ideal within this tradition.

Some aspects of the indirect responsibility model have been explored in the previous chapters. The facets of the indirect responsibility model that I will highlight in this section are:
- the responsibilities of the state;
- the position of the state;
- the orientation or rationality of the state;
- the organization of the relations with other fundamental institutions.

The responsibilities of the state

> 'The state can take on the most heterogeneous of problems and activities, as long as they are seen to spring from the general interest. For the aim of this bond is in fact none other than that of connectedness itself... to live together and to make action possible and to promote it.' (Van der Pot, et al., 1995: 135)

This quote from Van der Pot and Donner contains two crucial insights into the tasks of the state in a liberal democracy. The first, normative insight is that the state's tasks should be limited to public issues. The second, analytical insight is that, in modern-day society, a whole different range of issues can be classified as public. It is because of the inevitability of this heterogeneity that Mill (1848: 146, 151) points out the difficulty of systemizing and delimiting the state's tasks, something which economists from Smith (1776) onwards have attempted to do.

> 'The necessary functions of government ... (are) ... more multifarious than most people are at first aware of, and not capable of being circumscribed by those very definite lines of demarcation, which, in the inconsiderateness of popular discussions, it is often attempted to draw round them ...
> But enough has been said to show that the admitted functions of government embrace a much wider field that can easily be included within the ring-fence of any restrictive definition ...'.

Nevertheless, from an analytical point of view it is desirable to try to reach a systemization of the tasks that fall to the state. With regard to this issue it is possible follow up on a number of previous discussions. After all, within the liberal-democratic tradition there have been many inevitable controversies over the tasks and responsibilities of the state. An important conduit for the current discussion on this issue is the thesis of 'the neutrality of the state'. In principle this discussion could have served as a starting point in the search for a systemization or categorization of state tasks, were it not that the discussion on neutrality has little to do with our theme of controlling the market.

Remaining as close to this theme as possible, I have opted for an analytical systemization which relates to the three types of market imperfection from the previous chapter. These three types of market imperfection were market restrictions, market disadvantages and market imperfections (in the narrow sense). In our analysis we will relate the concept of 'market restriction' to economic goods only. As such, the concept refers to a *productive state task*.

The concept of market disadvantages refers to goods or aims which cannot be produced or achieved through the market because of the way in which the market operates. This corresponds with the notion of a *corrective task* for the state. Corrective action by the state implies in all cases that the state rearranges, adjusts or rules out the influence of market forces with an eye to achieving a goal, whether it be economic or non-economic.

The concept of market imperfections (in the narrow sense) covers situations in which the market as it exists does not meet the conditions of the perfect market. The task of realizing these conditions is the third main task of the state. Here we will refer to this as the *foundational task*. It can be described as creating the conditions by virtue of which the market can exist. In this regard it is useful to make a distinction between *constitutive* foundational tasks and *operational* foundational tasks. The former is concerned with the realization of general basic conditions for the free market, such as competition, freedom, private property and justice. The latter deals with the specific design of specific practices with a view to the realization of the free market. Here we are mainly concerned with the constitutive foundational task. Neoclassical economists, however, pay hardly any attention to this dimension of the foundational task. After all, they have the tendency to regard the market as a naturally occurring sphere. As a result, the constitutive foundational task has no place within their conceptual framework.

For a proper appreciation of this systemization of state tasks, two observations need to be made. First of all, it should be underlined that we are dealing with an analytical distinction. Actual state tasks often come about as the result of several considerations at the same time. The question of why roads are mainly under state control can be explained with reference to the productive state task. A road can be looked upon as an economic good that cannot be produced adequately on the market. After all, by its very nature a road seems to foreclose the possibility of perfect competition. However, one can also construct an argument against the privatization of the road system on account of the fact that in a liberal society people have to be able to move freely.

Secondly it is important to make a clear distinction between the *productive state task* and the *corrective state task*. The productive task of the state is limited in this study to economic goods which can be produced relatively independently from other social processes. The dependencies the state encounters in dealing with a productive task do not differ from those encountered by a producer of butter or shoes. Good examples of processes under the heading of productive state task are the production of electricity and water. In all those cases in which the realization of goods, aims or ideals are hindered by other social processes, we can speak of a corrective state task.

In other words, the state has a productive task when the realization of an economic good is obstructed by the presence of positive external effects (i.e. the impossibility of excluding those who do not pay for the benefit of an economic good). In cases where other problems of collective action are present (e.g. jointness of production) it is up to the state to act as a corrective body. In line with this conceptualization, environmental problems imply corrective intervention by the state in most cases. Lack of clean air, for example, is not due to the fact that clean air is impossible to produce but due to the fact that the air is being polluted.

The productive task

The state in a liberal democracy has a fairly extensive *economic* or *productive* function. Even though in a liberal democracy economic functions should be left to the market as much as possible, it is true to say that there are certain economic goods that the market cannot produce, or cannot produce as efficiently as the state. Economists refer to these goods as 'collective goods' (see Chapter 2) Variations in historical circumstances frustrate any efforts to give a complete list of such goods. Adam Smith (1776:720-730), for example, was of the opinion that canals were suitable for privatization but that roads were not. His argument was that a poorly maintained canal becomes unusable while a neglected road only becomes more difficult to negotiate. He therefore thought that the privatization of roads was a bad idea: the owner doesn't have enough incentive to keep a road in good condition, since customers have little choice but to keep using it. Nowadays this argument for the collective character of roads carries less weight. For one thing roads are far more plentiful than they used to be, which gives potential customers many more alternatives. Another reason is that means of transport have become more dependent on the quality of the road surface.

There are also some economic products that the liberal democracy wants to keep outside the market for normative reasons (see for example Dewey, 1980: 481 or Hughes, 1983; Walzer, 1980; Ulrich, 1993). This might be because the potential for achieving profit is seen to be too great or because people are afraid that the market power of the owner might become excessive. The fact that the Dutch government does not privatize the production and distribution of tap water has much to do with these normative considerations. The production of electricity provides another apt historical example of this second situation. Traditionally it has been kept outside the market in many Western countries because people were afraid that the power of the electricity producers would become too great and their profits too high (Hughes, 1983).

Of course, in almost every existing liberal-democratic society the state will always produce a number of economic goods that could just as easily be produced by a private party under the same circumstances and according to the same methods. This phenomenon covers situations like the production of ordinary economic goods by state companies like Renault's car factories in France. From a liberal-democratic perspective, society ought to reduce the state's production of such goods to a minimum. In more general terms, the liberal-democratic viewpoint demands that a constant critical watch be kept on whether the state is not producing a good without due reason.

The foundational task

The state derives its foundational task from the fact that the social order is not spontaneous in nature. Civil society and the market have to be set within a normative and judicial framework. They can only function well if clear norms and rules apply to these spheres and if individuals comply with them. The state does not have the exclusive right to produce these norms and rules but it does have an influential voice when it comes to the authoritative recognition thereof. That is to say, insofar as these rules are judicial rules, the state has an exclusive right to proclaim them. Since the market is dependent upon these judicial rules to such a high degree, the state has a very prominent foundational task in this regard.

Adam Smith (1776: Book 5) conceptualized the foundational task as 'the administration of justice'. I regard this conceptualization as being too limited. The administration of justice suggests that the rules and norms that apply to the market are known and unchangeable and that it is up to the state alone to settle disputes and enforce rules. 'The market' is not a naturally occurring phenomenon, however. It is a changeable institutional construction. The constitution of the market therefore involves more than just the administration of justice. It is an ongoing affair, in which rules have to be made and adjusted, differences settled and norms and rules enforced. The rise of the corporation neatly illustrates this process. The corporation only came into being as a legal entity relatively recently and is subject to a continual process of change. In one instance this might be because tax regulations are in need of improvement or in another because new situations arise, such as the separation of ownership and management which took place 100 years ago (Berle and Means, 1932; Stone, 1975).

The corrective task

Ancient Greek society is often seen as the cradle of our civilization. However, the classical Greek notion of democracy and the modern liberal-democratic interpretation are at odds with one another in some respects (Sartoni, 1987). Within the classical Greek idea of democracy, public life was a goal in itself. According to Greek sensibilities, leading a valuable and meaningful life was in many ways tantamount to leading a public life. This esteem for public life finds its continuation in the Republican tradition in political philosophy and as such has also influenced modern liberal-democratic thinking. Modern liberal-democratic thinking, however, is far more imbued with the contrary notion that the good life is lived in the private domain. Partly as a result of this 'shift towards the private' many liberal democrats share the view that, in an ideal society, the state should limit itself to its productive and foundational tasks.

Nonetheless, liberal democrats do allocate a third task to the state: the *corrective task*. The need for this additional task lies in the fact that constructing a perfect market is not the ultimate end of liberal democracy. Ultimately, liberal democracy is geared towards realizing freedom and equality for all. The ideal of the free market is subordinated to this ideal. This implies that the state has a prima facie task to correct the working of the market when it systematically stands in the way of realizing the ultimate goal.

Liberal democrats of course differ in their opinions on exactly which activities belong to the state's package of corrective tasks. According to Albert (1992) we should make a distinction between the Anglo-Saxon part of the Western world and the European continent. On the European mainland, there is a tendency to extend the range of corrective tasks much further than in the Anglo-Saxon world. An additional factor is the extent to which the deployability of the market and civil society changes with time. For example there used to be less need for the state to combat poverty at a time when the Christian church held society together to a significant degree. Such variations make it impossible to draw up a definitive list of corrective tasks.

It can be asserted, however, that the range of corrective tasks allocated to the state in all Western nations over the past 150 years has expanded considerably. To explain this phenomenon, I would like to introduce the term 'collective dimension'. The collective dimension of an activity or a process is the effect relevant to public issues, whether it be intentional or unintentional, predicted or unforeseen. The ways in which the actions of the business community affect the quality of the environment form part of their collective dimension. In modern-day society, this collective dimension is on the increase. As a result, the individual becomes ever more powerless in the face of an increasing number of issues (see Section 1.3). The other side of this effect, however, is the need for an ever increasing number of issues to be handled publicly. The burden that the state has to bear as a result of the expansion of the corrective package has become so great that there is now widespread support for the position that the state is overloaded (Huntingdon, 1975; Offe, 1979; Held, 1987).

In this study we mainly conceptualize environmental problems as market disadvantages and therefore part of the corrective task of the state. In the light of neoclassical theory, this demands a measure of explanation. After all, neoclassicists conceptualize environmental problems as market imperfections (in the narrow sense) and therefore as part of the foundational task. On the basis of the analysis of the neoclassical thinking from the previous chapter, however, I would argue that the neoclassical conceptualization of environmental problems rests on an inadequate conceptual framework. This conceptual framework confuses the functional and the empirical-historical concept of the market and assumes the idea of a perfect state.

In other words, environmental problems can only be treated as market imperfections (in the narrow sense) on the basis of highly unrealistic assumptions about the idea of a perfect market. In this study we will proceed on the basis of the empirical-historical market concept. Within this conceptualization it is, on the one hand, perfectly reasonable to see the environment as a non-economic good that deserves protection. On the other hand, within this conceptualization it makes little sense to put environmental problems in the same category as classical market imperfections (in the narrow sense), such as monopoly-forming. While monopoly-forming calls for an improvement in the way the market functions, environmental problems require corrective action (which might for example take the form of rearranging limiting conditions).

The position of the state

The position of the state in relation to other fundamental institutions within the liberal-democratic tradition changes according to the perspective we choose to adopt. From the normative perspective the state occupies a subordinate position. Liberal democrats think the market and in particular civil society are worth protecting, even if they do not fulfil certain responsibilities or do not do so as well as might be hoped. These fundamental institutions are necessary to provide a context for 'the good life' and herein lies their value. The liberal democrats are much more ambivalent when it comes to their normative evaluation of the state. On the one hand, they do see that the state has some inherent value as the main locus of politics and democracy, but on the other hand, the 'shift towards the private' has led them to conceptualize the state on a purely instrumental basis. Sections of the state which are unable to prove their social function therefore have no right to exist.

If we switch from the normative perspective to the managerial one, then suddenly we find the state enjoying full status within the liberal-democratic tradition. After all, the state has both an important task in the constitution of society and in the handling of market imperfections. Given the weight of the burden of this managerial role, liberal-democrats are of the opinion that the state should also be given a prime position when it comes to power. If the state is to acquit itself of its responsibilities, then it should be in a position to alter the framework of society and/or make corrective interventions in social processes. This it can only accomplish from a position of formidable power.

To a certain extent, the position of power occupied by the state is based on pure or actual punitive power. The state can and should have the power to impose its will on its citizens and other actors. However, a society in which the state frequently and structurally has to resort to this power base in order to control its citizens can no longer be called a liberal democracy. In a liberal-democratic society, the state exudes authority. This means that the citizens accept the state as a legitimate power (Weber, 1921; Habermas, 1992). The fact that power is exercised legitimately means that citizens usually obey the rules without the authorities having to resort to the threat of force. The regulations have the support of the citizenry.

The orientation (rationality) of the state

According to liberal-democratic thinking, the position of power enjoyed by the state is based on its legitimacy. This legitimacy, in its turn, depends upon the extent to which the state can enforce its own logic or rationality. Research into state bureaucracy in the 1950s and 1960s tended to interpret this rationality as efficiency (Gouldner, 1954; see also Albrow, 1970). From a liberal-democratic point of view, however, this is not entirely accurate. Efficiency is only one criterion by which to measure the functionality of the state and it is certainly not the most important. Max Weber (1921) provides a simple illustration of this point. Weber identifies one of the tasks of the state as ensuring that the market functions efficiently. Above all this means that the state has to be reliable and punctual. These criteria for the functioning of the state are therefore much more important than efficiency itself.

The environmental philosopher J.S. Dryzek (1987: 25) has proposed that the rationality of the state be contingent on the extent to which it is capable of realizing a sustainable society. This criterion is too reductionist, however. By orienting his definition towards sustainability, Dryzek denies that in a liberal democracy the state has to take into account other values, such as freedom and equality. What is more, Dryzek fails to consider the fact that, in a liberal democracy, the way in which the state achieves its targets is at least as important as whether it achieves its targets. Ideas pertaining to the *Rechtsstaat* and control by limiting conditions (*Dauerlenkung*) cannot simply be set aside.

For these reasons, I regard it as fruitless and misguided to try to capture the rationality of the state in a single criterion, whether it be efficiency, sustainability or another value. Claus Offe (1985) confirms this position. He points out the divergent nature of the rationality of the state within a liberal democracy. Any judgement on how adequately the state functions can only be made on the basis of a number of criteria which are difficult to bring together. Building on Offe's line of argument, we can divide the rationality of the state in a liberal democracy into four - not always equally harmonious - components: effectiveness, *Rechtsstaat*, democracy and prudence. Effectiveness in this context should be understood as the extent to which the state achieves the goals set, such as its ability to take corrective action and/or set sound limiting conditions. In the following section I will discuss the component of the *Rechtsstaat*. Here I would like to briefly deal with the concepts of democracy and prudence.

Democracy as the sovereignty of the people

The content of the concept 'democracy' is hotly debated within the liberal-democratic tradition. Precisely because this concept is so essential to the liberal-democratic order, this level of controversy should not be surprising: anyone who is able to set themselves up as the true guardian of democracy occupies an almost unassailable position in any public discussion. Yet, among the liberal democrats there is a form of 'minimal consensus' according to which democracy means at any rate that the citizens are sovereign.

'Sovereignty of the people' can be roughly explained as the idea that in a democracy 'the citizens rule' (compare Dahl, 1986: 18-19; Lively, 1975: 30 or Sartoni, 1987: Part I, 28). More precisely, sovereignty of the people means that the citizens form an autonomous political-administrative association which has the right to make changes to the law and which in that sense is above the law. Sovereignty of the people also implies that the citizens exercise the highest authority: they are not responsible to anyone, while every other political-administrative organ within the association is answerable to them (Hobbes, 1651).

In a large, modern society, the idea that 'the citizens rule' can only be materialized symbolically (see also Van der Pot et al., 1995: 123 and 393-399). The actual meaning of the idea can therefore best be read by looking at the way in which this notion has been translated in institutional terms. The notion of sovereignty of the people then comes down to the idea that, in a democracy, the citizens directly or indirectly elect government leaders, as well as their political representatives. In other words: in a democracy the citizens decide directly or indirectly who takes up the reins of executive and legislative power. Rödel and his fellow writers (1989: 59) express this beautifully with the notion that in a democracy no one occupies the position of power, at least not permanently: *'The choice of a consistently plural symbolisation of the people means that the position of sovereign power tends to be empty'*.

The fact that the liberal democracy attaches a great deal of importance to the sovereignty of the people can be measured by such indicators as the internal organization of the state. The hierarchy and the strict dividing line between the administrative sector (non-elected) and the political sector (elected) are two important organizational principles. Both are also of great importance for democracy. The citizens can get by with only electing the representatives at the top of the state (the norm in the European situation) as long as the organization is strictly hierarchical. The strict dividing line between politicians and civil servants is also important for this reason. Only if these roles are kept strictly separate will it remain clear who leads and decides, and who carries out these decisions.

Democracy as consultation

Many liberal democrats are of the opinion that indirect representation fails to do justice to the democratic ideal. A properly arranged democracy is also characterized by all manner of guarantees for participation by citizens in the policy process. I call the total sum of these elements 'democracy as consultation'. Democracy as consultation gives those who are directly affected by a given policy the opportunity to make their voices heard and to try to influence the process of policy-making (Van Wijk and Konijnebelt, 1991: 45-52). According to most liberal democrats, the increased complexity and specialization of modern society make the necessity and the desirability of democracy as consultation more and more pressing. Consultation gives citizens new possibilities to influence political-administrative processes now that the distance between citizen and government is growing ever larger out of necessity. R. Paehlke (1989) is an example of an author who, within the context of sustainability, makes a forceful plea for democracy as consultation.

Modern-day advocates of participatory democracy and/or interactive policy-making appear to regard democracy as consultation as a new interpretation of the notion of the sovereignty of the people. In my view it is better to see it as *compensating* for the evaporation of democracy as the people's sovereignty (see Section 4.2). An argument to this end is that democracy as consultation is often advocated as a response to the growth of administrative tasks which need to be carried out by the state. Administrative tasks usually involve the freedoms and opportunities accorded to specific individuals. Democracy as consultation gives individuals in this situation the chance to resist when the state threatens to influence their specific opportunities or rights (for a critical discussion see Mashaw, 1985).

The scope for democracy as consultation is to some extent determined by law. A Dutch example is the opportunity that citizens have to state their case when it comes to major planning decisions. In many Western democracies, however, the actual significance of democracy as consultation goes far beyond the legally established possibilities for consultation. In the Netherlands and the United States, it is common in many policy areas for the views of interest groups to be taken into account at administrative and/or political level before the policy is formally presented to parliament for approval (see for example Weale, 1989 or Yeager, 1991). We can interpret this process as a form of consultation, even though in many cases it cannot be regarded as wholly democratic, since far more attention is paid to some groups than to others.

Democracy as culture

The question of whether democracy constitutes more than this corpus of procedural regulations regarding the electing and influencing of government officials is one that divides liberal-democratic thinking. Some see democracy as a way to elect (and exert influence on) administrators and nothing more or less than that (Schumpeter, 1944). Others feel that the idea of democracy cannot be reduced to rules and procedures. In their view, democracy means striving towards a society of liberal-democratic design, where everyone enjoys certain rights and freedoms: democracy as a way of life, a culture. The procedures surrounding the choice of administrators only represent a part of this.

This difference in insight was especially prominent in the debates on freedom and equality which took place in the 1960s and 1970s. From a cultural or content-oriented vision of democracy, a direct link can be made between the ideal of democracy and a degree of positive freedom for all. From a procedural point of view this link is not so apparent. In the present-day debate on democracy and sustainability, this difference in insight is relevant to the discussion on the extent to which the state can and/or should distance itself from the opinion of the majority. Someone who sees democracy as pursuing the implementation of specific values will be more inclined to conclude that the government may deviate from the views of the majority than someone with a procedural approach to democracy. From the cultural perspective it is, after all, possible to regard the pursuit of sustainability as democratic, even if the majority opposes it.

In my opinion, the cultural concept of democracy is more fitting than the procedural. The procedural vision cannot serve as a basis for arguments concerning what democracy is actually good for. Democracy from a procedural viewpoint, lacks justification and meaning. Any justification necessarily leads to the question of which values are important, and that is exactly the leap that the procedural approach does not want to make.

On the trail of the cultural vision of democracy many of today's thinkers are searching for a new interpretation of the concept of sovereignty of the people. They describe the people's sovereignty as a form of government in which the state is led as much as possible by the sound arguments which crystallize in the discussions which citizens have with one another in civil society (Cohen and Arato, 1992; Habermas, 1981 and 1992).

Prudence

When it comes to defining its tasks and setting itself concrete goals, the state always seems to be caught between the devil and the deep blue sea. On the one hand there is the world of ideals and what is desirable (freedom, equality and sustainability), while on the other hand lies the world of actual possibility, of technological capability, the economic situation as it stands, the balance of power and other unruly practicalities.

The notion that the state in a liberal democracy should be prudent, reflects this tension. A prudent administration will succeed in creating appealing trade-offs between these opposing worlds. A prudent administration allows itself to be inspired by liberal-democratic ideals but does not lose sight of reality. It simultaneously draws on an idealistic component and a strategic component. The latter should not be seen as a set of objectives which have been entirely stripped of normativity. Behind this demand lie values such as the search for stability, order, peace and piecemeal engineering.

Various authors have pointed out that in many Western societies the idealistic component is coming under particularly intense pressure (see for example Habermas, 1981 and 1992; Dubbink, 1998a). The inadequate way in which the ideal of sustainability has been worked out in practice provides ample illustration of this. With this in mind I would like to reformulate the prudence criterion. From this point on, prudence will be seen as the extent to which the state succeeds in doing right by the idealistic component of government.

The organization of relations with other fundamental institutions

Modern society is a differentiated society. The liberal-democratic tradition views this differentiation (between market, state and civil society) as desirable (Cohen and Arato, 1992). Differentiation allows each sphere to develop as much as possible in accordance with its own rationality. Liberal democrats see this as beneficial to the realization of liberal-democratic values, as opposed to dedifferentiation, which can quickly undermine these ideals. The efficiency of the market, for example can be rapidly undermined by political-administrative interference in the market process. It is therefore vital that politicians intervene in the market as little as possible and limit themselves to controlling the market by means of limiting conditions if required (see Chapter 2). The other way round, the involvement of market parties in the political process soon leads to whispered dealing in backrooms, domination by short-term interests, favouritism and an atmosphere of 'you scratch my back and I'll scratch yours' (see for example Ankersmit, 1995; Cohen, Rogers and Wright, 1995; Cohen and Arato, 1992; Freeman, Kneese and Haveman, 1973). All these factors erode the prudence of the state. The market should keep out of state affairs.

A third reason why liberal democracy attaches importance to differentiation has to do with the notion of pluralism. It prevents a situation in which all centres of power cluster together. From a liberal-democratic point of view this is important. One of the liberal-democratic nightmares is that of a society in which the individual cowers naked and unprotected in the face of a single bastion of power. Accordingly every liberal-democratic society ought to have several centres of power.

Typical features of the indirect responsibility model

The indirect responsibility model comprises one possible interpretation of liberal-democratic thinking at the level of social organization. How can this model be characterized in the light of the above? A typical aspect of the indirect responsibility model is that its advocates are inclined to allocate public issues exclusively to the state. 'The state' and 'the public issue' are almost synonymous terms within the indirect responsibility model. The liberal-democratic position that the state should concern itself only with public issues is also interpreted back to front as it were. Public issues are regarded as state property. This view transforms the market and civil society into fundamental institutions in which actors only have a very limited public responsibility. In the market this responsibility is limited to observing the limits set by law and common decency.

When it comes to the position of the state, the view of the indirect responsibility model is characterized by a significant measure of trust in the potential of this fundamental institution. This mental model pays little attention to the possible limitations of the state. Where it is concerned there are no tasks which are impossible for the state to carry out.

In relation to democracy, it is striking how easily the indirect responsibility model has reconciled itself to the idea that the sovereignty of the people can only have symbolic meaning in modern society. Advocates of the indirect responsibility model accept representative democracy and its dividing line between citizens and administrators. One marked consequence of this is that within this model little attention is paid to the position of civil society when it comes to public issues. The citizens only become periodically relevant in their role as voters.

When it comes to the links between the fundamental institutions, we can refer to the advocates of the indirect responsibility model as the true defenders of the liberal-democratic faith. They set a great deal of store by maintaining clear dividing lines between the fundamental institutions. A no-man's land, an impassable minefield; these are the metaphors which spring to mind when trying to describe the indirect responsibility model's view of relations between the state and the other fundamental institutions. Only two paths snake across this tortured landscape. On the 'input side' of the state there are the democratic procedures which bridge the gap towards the citizens and civil society. On the 'output side' of the state there are the general laws and rules that the state uses to govern society, which constitute control by limiting conditions.

3.3 RIVAL TENDENCIES

My ideal-typical reconstruction of the indirect responsibility model has been much more extensive than the sketches to be found in the administrative literature. In many cases the reader has to make do with a few catchwords tossed around by the author such as 'bureaucratic' or 'top-down'. However, in my view the administrative claim that the indirect responsibility model has been cast off can only be proved sufficiently if this mental model has first been set out clearly and once the intimate relationship between the indirect responsibility model and the liberal-democratic tradition as a whole has been teased out.

While it is true to say that my representation of the indirect responsibility model is more extensive than the picture to emerge from the administrative literature, so far it has not been significantly different. In this section, I would like to further refine my sketch of the indirect responsibility model and, in doing so, I will leave administrative thinking behind me. The point is that administrative considerations tend to ignore the normative aspects of thinking about the social organization of society. Administrative theorists tend to reduce the question of organizing modern liberal-democratic society to a strategic or technical matter. I think this reduction is based on a category mistake. Modern society cannot and should not be organized as a special-purpose tool, but in accordance with the norms and values that the liberal-democratic tradition stands for. This implies that there are sometimes aspects of society that will be organized in a hopelessly expensive, awkward or clumsy way.

What makes this normative dimension of special importance is that liberal-democratic thought contains ambivalences and tensions, that stem from conflicting normative ideals. This implies that *any* liberal-democratic mental model of social organization will be torn between two (or more) worlds. The internal agitation within the liberal-democratic tradition itself renders the construction of a consistent mental model of social organization impossible. No mental model of social organization can be perfect. This conclusion has important implications for the evaluation of liberal-democratic mental models. When criticizing one specific mental model, one always has to take into account that any alternative will have its own problems and deficits. Ignoring that, as administrative theorists tend to do, means taking the easy way out.

To sum up, one might say that I doubt whether the administrative theorists capture the essence of the problems with the indirect responsibility model. I think these problems are mainly normative in nature and related to tensions within the liberal-democratic tradition. We might describe them in the terms of classical tragedy: they represent a *'conflict of true values with each other'*. In this section I will try to articulate the tensions and the normative ideals that cause them.

A powerful and curbed state

The first clash of rival tendencies within liberal-democratic thinking is in the simultaneous pursuit of a powerful and a curbed state. That the state should be powerful is not so much an ideal as an inescapable fact within liberal-democratic thinking. In order for the state to carry out its huge range of tasks, it must be equipped with a battery of far-reaching competences. These include competences relating to the drawing up of laws and regulations, enforcement of existing laws, organization of security, external and otherwise, the dispensation of justice and the collecting of taxes.

Yet this view alone fails to sum up the liberal-democratic view of the power of the state. Liberal democracy also tries to limit the power of the state, thereby rendering it something of a fettered giant. Behind this contradictory tendency lies the realization - strong among liberal democrats - that the power granted to the state in order to structure society in accordance with the liberal-democratic cause can also be turned against the citizens and their autonomy. Classically, this realization is expressed in the rhetorical question *'who guards the guardian?'*, and reflects a fear of corruption, partisanship, arbitrariness (Bobbio, 1987), mob rule (Sabine, 1973: 95-114) or a coup by a military elite or an elite allied to the military.

Present-day liberal-democrat theoreticians have added one more threat to this traditional list, one which has become the focus of increasing concern: the danger that liberal-democratic society - either intentionally or unintentionally, in plain sight or largely unnoticed - may turn into a totalitarian society (Tocqueville, s.a.; Talmon, 1952; Cohen and Arato, 1992: Pels, 1992). Many thinkers warn that we should not underestimate this danger of totalitarianism. We should also be well aware that this threat has taken on a different guise. If the modern liberal democracy goes off the rails, it will probably not be due to the machinations of a select few. It is much more likely that well-intentioned efforts to secure positive freedom for all will metamorphose into their opposite under the pressure of the increasing collective dimension. It is exactly because of this paradox that distrust of the state should be kept alive, and the citizens in the liberal democracy should never be lulled into a false sense of security by the well-intentioned chorus of administrators, politicians, civil servants or other friendly monsters whose awkward embrace can stifle freedom. This fear that the state may acquire too much power, and in particular the fear that, on the rebound, it can go from being a loyal servant to public enemy number one, motivates liberal democrats to emphasize the need to limit the power of the state. The determined efforts to clothe the state in power are balanced out by equally determined efforts to limit its power (see for example Barber, 1990 or MacPherson, 1977). Attempts to give form to this curbing of the state can be seen in a large number of liberal-democratic ordering principles. In order to show how seriously liberal democracy takes this task of limiting the power of the state, I will now proceed to underline five of these tenets.

Five safeguards against the power of the state

(1) The first liberal-democratic principle aimed at fencing off the power of the state is the distinction between public and private issues, already examined in Section 1.3. This distinction delimits the kind of issues that the state can legitimately take as a reason to (re)structure institutions or otherwise interfere with the lives of citizens. The state must show that a public issue is at stake.

(2) The second principle has also been examined previously. It is the distinction between the private and the public domain. This distinction delineates a specific sphere of action within society in which the state cannot legitimately coerce or otherwise control individuals. I have already indicated (in Section 1.3) that these domains should not be interpreted as spatial areas. They delineate specific domains of choice. Another qualification I would like to stress is that the borders of the private domain cannot be set down for all time. Sometimes historical circumstances and new insights impel the redrawing of the border between the private and the public domain. For example in the 19th century children almost completely belonged to the private sphere and were thus at mercy of their parents (i.e. their father). Nowadays the rights and liberties of parents with regard to their children are limited, because we feel that children sometimes need to be protected against their parents. Still, the fact that the border between the public and the private is historically indeterminate does not make the distinction practically obsolete. If the state wants to act in a way that goes against the principle that the private sphere is not to be invaded, it must legitimize its actions on two levels: it must justify both the action itself and the proposition that the border between the public and the private needs to be redrawn.

(3) The third liberal-democratic principle aimed at reining in the state is the primacy of self-governance or self-organization (Mill, 1848: 312-314; De Haan, 1993). According to this principle, the state is always the last resort when dealing with public issues. The state may only come into action if the citizens are unable or unwilling to deal with a public issue themselves. If the citizens are able and willing, the state should yield or even facilitate the citizens' self-organizing initiatives. One qualification should be made with regard to this principle. The principle of self-governance is very dear to some liberal democrats, for example J.S. Mill (1848 and 1861). However, not all liberal-democrats explicitly embrace this principle. Indeed, its value and importance are hardly touched upon by advocates of the indirect responsibility model.

(4) The liberal-democratic tradition has also formulated other principles which restrict the way in which the state can operate. The most important of these principles is that the state in all its forms - as a private party purchasing stationery, as an administrative power, as a lawmaker and as a judicial power - is subject to laws, both written and unwritten. Insofar as this principle relates to the state's important role as an administrative organ, it is often expressed in the concept of the *Rechtsstaat*. Van Wijk and Konijnebelt (1991) attribute a number of concepts to this ideal. 'Legal control' and 'lawfulness' are two of the most essential. The latter refers to the fact that the state only has at its disposal the powers explicitly granted to it by law.

(5) Even this range of restrictions is not enough to satisfy the liberal-democratic tradition. It has also developed a number of principles aimed at frustrating the formation and accumulation of power *within* the state. The doctrine of the *trias politica* or the division of the competences should be understood in this context. This doctrine states that legislative, executive and judicial competences should be kept strictly separate from each other to prevent the concentration of power. It is

evidenced, for example, by the strict distinction between the judiciary and the executive power which is a feature of Western democracies.

Another such principle is the doctrine of checks and balances. This is not so much aimed at the separation of competences as at the distribution of power within a sphere of competence. One tried and tested method is to give a certain power jointly to two opposing bodies. A Dutch example is the division of legislative power between the government and parliament. The United States also makes use of this principle when it comes to legislative power. Congress passes federal laws but the president has to sign them in order for them to become law.

An ideal which should also be mentioned in this context, is the pursuit of institutional pluralism. This concept can be interpreted as working towards a society in which the power of the state is kept in check by a lively civil society and a vital market, both of which bristle with organizations possessing a measure of power and autonomy (Dahl, 1986).

Fighting for and against the planning of the market

Thinking with regard to planning the market process forms the second contradictory tendency within the liberal-democratic tradition. Liberal democracy is strongly opposed to planning the market. Key concepts of liberal-democratic thinking are after all 'the decentralized order' and 'the freedom of the individual'. This set of ideas is fundamentally opposed to the idea of an economic order in which actors have their objectives and/or working methods imposed upon them from above. The planned economy can without question be referred to as the antithesis of the liberal-democratic ideal.

> '*Government by rules, whose main purpose is to inform the individual what is his sphere of responsibility within which he must shape his own life*, differs fundamentally from *government by orders which impose specific duties*.' (Hayek, 1980b: 18)

Nevertheless, whoever interprets the liberal democracy as the antithesis of planning forgets that this tradition also contains an equally fundamental desire for planning, albeit an unintentional one. After all, liberal democracy wants to achieve a well-organized society, a social order in which at least actual negative freedom for all is realized. The market is ultimately subject to this normative goal and not elevated above all criticism. If the market structurally obstructs the aims of liberal democracy then it can and should be controlled. Liberal-democratic thinking makes no mistake about this. As Dahrendorf (1966: 11) clearly states:

> '*Within the framework of our discussion, the following aspect of this fact is of particular interest: market policies clearly presuppose planning decisions, if they are not to remain the ideology of the systematic favouring of those who are already in a position to take part in the processes of exchange and competition*'.

This rival tendency is of course highly relevant to this study. On the one hand, liberal democracy strongly resists planning. On the other hand, from the perspective of sustainability there is a continual unintentional drive towards controlling the market more exactly. The risk that this effort to control the market *de facto* will end up as an attempt to plan the market should be seen as real. The feasibility of this idea is underlined by liberal-democratic theory on totalitarianism. Authors in this tradition argue that society is distancing itself more and more from the ideal of control by limiting conditions (Tocqueville, s.a.; Ankersmit, 1992; Foucault, 1977). The state no longer limits itself to drawing up general rules. Instead it increasingly sets out to shape the conduct of its citizens in specific ways. This is typical of a planning or tutor state.

In this respect, recent Dutch legislation pertaining to the health and welfare of animals (Handelingen der Tweede Kamer, 1981, 1985 and 1986) makes for interesting reading. In the chapter on welfare, the legislator has opted to ban all actions towards animals apart from those which are explicitly permitted. It would be difficult to imagine a clearer illustration of tutor state.

Also relevant here is the theory developed by Christopher Stone. Stone (1977: 31) observes that the present-day state is turning its attention more and more towards the influencing of what he describes as *qualifiedly disfavored conduct*. This type of behaviour occupies a position between two poles. The first of these is *absolutely disfavored conduct*. Stone defines this as conduct that is forbidden in all contexts and in all manners. Murder is a clear example. The other pole - that Stone does not identify explicitly - is 'approved conduct'. This kind of conduct can be described as behaviour that is unquestionably permissible. Qualifiedly disfavored conduct occupies the middle ground between these two extremes: it is behaviour that is desirable and permissible in principle but which has certain undesirable side effects. In order to counter these side effects, qualifiedly disfavored conduct needs to be designed in a certain way.

The concept of qualifiedly disfavoured conduct shows precisely where the trouble with planning lies. On the one hand, liberal-democratic thinking is against planning. On the other hand, environmental problems confront the liberal democracy with a growing collective dimension. The growth of the collective dimension means that much of the conduct on the market can be regarded as qualifiedly disfavored conduct. Although it is desirable, it also tends to have negative consequences. Accordingly it should not be prohibited but steered in the right direction. The manufacture of product X is permitted, but only in such and such a way in order to prevent the occurrence of undesirable effect P, as well as undesirable effects Q to Z.

The control of the paint chain is a good example of this development (Van Vliet, 1994). The production of paint is a profit-generating economic activity which places a burden on the environment. As such it would be out of the question simply to allow it with no conditions attached. Straightforward prohibition is equally out of the question, not least because paint protects wooden objects and is therefore necessary to a sustainable society. For a state determined to restrict itself to limiting conditions and which is not prepared to take ecological risks, there is not much alternative but to ban many kinds of paint or raw materials. However, such a policy would be far too rigorous. After all, the hazardous nature of paint is strongly related to its use. The use of acrylic paint, for example is only ecologically responsible if paint residue does not enter the environment. The specific ways in which paint is used and handled therefore have to be more stringently controlled. Rules and legislation to that effect, however, stray into in the lane of planning.

Typical features of the indirect responsibility model

The rival tendencies are intrinsically intertwined with the indirect responsibility model. They can be recognized in the fact that the model is not only about empowering the state. It is also concerned with curbing state power. The distinction between the public and the private, for example, looms large in the model, as does the *Rechtsstaat* and ideas about limiting the powers within the state. In fact the only principle ignored in the model is the primacy of self-organization. Furthermore, we should interpret the model's emphasis on control by limiting conditions as a tribute to the ideals of liberal democracy. Control by limiting conditions is an attempt to preserve a fragile balance between the unintended pursuit of planning and the explicit efforts to oppose it.

3.4 THE INDIRECT RESPONSIBILITY MODEL CRITICIZED

Most administrative theorists do not hold the indirect responsibility model in very high esteem. Open any book on administration theory, especially a Dutch one, and you won't have to read very far before the critical salvos start flying. Administrative theorists see the indirect responsibility model as a thing of the past, a worn out mental model. Today's society needs to find itself a new model as quickly as possible. According to the administrative theorists, society will not be able to handle public issues satisfactorily until this new mental model is found.

In this section I will give a comprehensive account of the administrative theorists' critique on the indirect responsibility model. Once again, I will take an ideal-typical approach in my presentation of this critique. The historically accurate rendition of administrative arguments has therefore been sacrificed to the aim of achieving a rigorous representation in systematic terms. One implication of this choice is that no coverage is given to arguments which are clearly incorrect. In addition I have, where possible, sought to reinforce the positions adopted by the administrative theorists with arguments from outside the administrative field. I have also organized the critique in four main arguments listed below, each of which is represented by a core statement. In the eyes of administrative theorists, the advocates of the indirect responsibility model fail to fully grasp:

- the extent to which the state is overloaded by an increasing number of tasks and responsibilities;
- that within the present-day context, the state is often unable to exercise its powers to the full;
- that the legitimacy of the state in contemporary society is under considerable pressure;
- that the powers accorded to the state are not or are no longer geared towards the kind of management that is needed nowadays.

These four main arguments will be explained in greater detail in the rest of this section.

An overload of tasks and responsibilities

The first argument that administrative theorists use against the indirect responsibility model is that the potential of the state to manage or govern social processes has turned out to be a good deal less extensive than many people in the 20th century hoped (Teubner and Willke, 1984; Hood, 1976; Tjeenk Willink, 1984). Now, at the dawn of a new century, it is clear that the state simply cannot cope with the tasks assigned to it (Keane, 1988b; WRR, 1992). The main cause of this problem would appear to be the spectacular growth in the range of tasks assigned to the state.

No one would presume to dispute the fact that such an increase has taken place. However, commentators are less in agreement when it comes to identifying the nature of the processes that underlie this development. Some see a connection with the increasing demands and wishes of the general public, as well as a growing tendency towards opportunism and greed among the ordinary citizens (Rose and Peters, 1977).

In my opinion, the growth in the collective dimension of actions and processes offers a better explanation for the wider variety of tasks allocated to the state (see also De Swaan, 1989). As a result of this growth, the individual in the 19th and 20th centuries has come to occupy a powerless and helpless position in the face of undesirable social developments. More and more issues have become collective and public in nature. Accordingly, the help of the state has also been enlisted more often. It has proved impossible for the state to keep pace with this growth of the collective dimension, notwithstanding the fact that the 20th century saw spectacular growth in both the civil service workforce and central government budgets of all Western countries (Ringeling, 1984). However, the odds were stacked against the administrators, since it was not only a question of finding ways to deal with an increasing *quantity* of tasks. There was also a *qualitative* aspect to this growth. Public issues often turned out to be interconnected and it is these links which have proved to be the undoing of the state.

The flatfish fisheries sector in the Netherlands presents a telling example of this phenomenon (Dubbink and Van Vliet, 1996; Dubbink and Van Vliet, 1998 or Dubbink and Van Vliet, 2000). With the means available to fishermen at the beginning of the 20th century, it was physically impossible for them to overfish the North Sea. They simply did not have the technology. What is more, the market for a product with as short a shelf-life as fish was limited. However, with the passage of time, technological innovation has completely altered this situation. Shipbuilding and fishing techniques have reached unimagined heights of technological and logistical expertise, while inventions such as electric refrigeration and rapid transportation have brought about exponential increases in the market for fish. Technological innovation has created a situation in which we are confronted with the danger of overfishing the North Sea on an annual basis. Furthermore, it is becoming all too clear that these intensive fishing practices are also taking their toll on commercially unattractive varieties of fish and other forms of life dependent on the sea, such as fish eaters like birds.

As a result of these developments, the state finds itself burdened with two additional tasks: preserving the commercially relevant flatfish stocks and protecting the salt-water ecosystems. Both tasks constitute a heavy load, even more so when we consider the complex ways in which they are tied up with the responsibility for ensuring economic prosperity and the social task of maintaining the livelihood of vulnerable communities in remote areas.

Powers cannot be exercised effectively

According to the indirect responsibility model, the state enjoys a position of formidable power. It can make laws and has at its disposal an apparatus for enforcing compliance with these laws. Accordingly the question of whether this position of power is sufficient in order to control the market never arises. Unjustly so, argue the administrative theorists. According to them this position of power has come to be undermined in modern-day society. The three remaining key arguments all have to do with this notion of the crumbling power of the state. Below we will discuss the first of these arguments: that the state frequently appears unable to utilize its powers effectively.

The administrative theorists point out that, to some extent, the powers of the state in modern-day society only exist on paper. In practice, the state is regularly confronted with situations in which it cannot put its powers into action. This is most definitely true with regard to the market. Market parties turn out to be capable of curbing the state in the exercising of its powers. They are able to bring considerable power to bear (Bowles and Gintis, 1986; Breiner, 1995; or Galbraith, 1991).

In the case of sustainability, the chance that actors in the market will oppose the state's effort to control the market by restrictive laws is relatively high (see Mayntz, 1978; Peterse, 1990; Yeager, 1991). Sustainability often involves making new demands on businesses. This process always entails the risk of lower profits. Businesses might also fear for their competitive position. As a result, many businessmen set out to resist measures aimed at sustainability. In the context of an order based on the indirect responsibility model, one should not be too quick to call such behaviour irrational or immoral. In such a regime parties in the market are expected to act as economic agents with limited public responsibility. Opposition to the state might therefore be described as rational in many cases or inevitable at the very least.

From the perspective of the indirect responsibility model, the presence of forces in the market that obstruct the workings of the state is disconcerting in itself. What makes the situation even more serious is that the power of market parties is so much greater than the power of other groups in society. Lindblom (1977: 170 et seq.) in this context speaks of 'the privileged position of businesses'. These privileges make themselves felt in all phases of the policy process: preparation and development of policy, decision-making, implementation of regulations and enforcement . A classic study of the power of businesses in the implementation phase is Renate Mayntz's study of the implementation of environmental legislation in the Federal Republic of Germany. Mayntz (1978) shows that federal environmental officials were continually forced to negotiate about the implementation of federal legal norms. This was due partly to the tendency of local government to take the side of the business community in order to safeguard jobs in their region.

Another important text in this field is *The Limits of Law* by Peter Yeager (1991). In this study, Yeager gives an account of the significant influence which businesses exerted on the drawing up of a federal law to protect surface water from pollution in the United States. To this end, corporate representatives did not even have to resort to secret lobbying, since business representatives formed the majority on the advisory committee set up by the United States government to assess the new environmental legislation.

Where do businessmen derive their power over the state from? One basic source of power is the market structure, and in particular the fact that the modern market is a market made up of organizations (Scott, 1981; Coleman, 1982; De Jong, 1985). The spectacular development of the corporation as a legal entity since the 18th century has made it possible to concentrate factors of production in considerable quantities over the course of time. Even in a market where the lashes of market discipline are clearly felt, corporations are able to amass market power by virtue of their size.

The power of market parties is further enhanced by the fact that economic prosperity is considered an important end within liberal democratic societies. Given this end, the state is always required to engage in a process of give and take, however important the pursuit of other values such as sustainability may appear. In other words, the power of market parties is reinforced by the fact that the state always has to balance the strategic and idealistic components of government. Measures which represent a possible danger to the stability of the market and/or society are rejected out of hand. Instability is a risk that the state cannot afford to take. It might for example lead to the drying up of the tax flow which the state needs in order to carry out its tasks (Offe, 1979).

This is not to say that the power of businesses is always a conscious and collective power. For example when measures are taken which the business community sees as entailing certain risks, this has an effect on every corporation's and individual businessman's willingness to invest. Without consultation and maybe even unintentionally, this also puts pressure on the state.

The need to govern in the spirit of give and take itself reinforces corporate power over the state still further. This type of government implies that the state must seek to control the market with precision and attention to subtleties. After all, such a well-considered approach will bring the greatest likelihood of achieving the best compromise. But this attention to detail also has a negative consequence. It makes the state dependent on sound knowledge and information. Market parties have far greater access to such information than the state (see also Weber, 1921:574).

The design of many pieces of environmental legislation reflects this dependency. In many environmental laws, the norms a business has to live up to are related to the context. One example of this is the ALARA principle, used in the Netherlands to regulate the emission of certain gases. ALARA stands for 'As Low As Reasonably Achievable'. The ALARA principle demands of businesses that they reduce their emissions to as low a level as possible, given the social and economic conditions and the sophistication of the available technology. A principle of this kind binds the effectiveness of the state to a great extent to businesses' willingness to cooperate, since the market actors are the only ones capable of providing the necessary information at an acceptable price (Mayntz, 1978).

The costs involved in the enforcement of environmental laws are the last source of power I would like to mention. These costs are often very high. This is partly due to the fact that environmental offences fall within the category described by Jerome Skolnick (1968: 629-630) as 'crimes without victim'. In a crime without victim, our exemplary idea of a crime no longer applies:

> 'Ordinarily, crime and the criminal are visualized as an involuntary interaction containing an element of assault. The criminal is a man who strikes at a victim and takes off ill-gotten gains. ... The prototypical criminal is a robber, a man who violates persons and property. ... Such a conception of criminality, however, also implies a conception of the role of police in society. Police are perceived as protectors of individuals who have been wronged by other individuals, and the police represent an organized social institution for coming to the aid of victimized individuals. ... By contrast, in victimless crime the collectivity is the victim rather than the individual. This fact leads to a number of unanticipated consequences that tend to undermine enforcement.'

The most important unforeseen consequence in this context is the absence of a plaintiff in the case of environmental offences. The victim cannot speak up and parties willing to do so on the victim's behalf - such as environmental groups - often lack the necessary antenna to register the offence in the first place.

> 'The offending event must somehow be observed and reported before sanctioning processes can be invoked. The potential efficiency of a control system, varies, therefore, with its capacity to receive or observe reports of offense.' (Skolnick, 1968: 630)

The problems relating to the punishment of environmental offences once again force the state into the arms of the business community. Businesses are often asked to do their share when it comes to enforcement. It is obvious that this way of working leads to dependency (Mayntz, 1978).

It is important to realize that the market power which hinders the state in its efforts cannot always be traced back to the unwillingness of businesses. Market parties are subject to the discipline of the market. Insofar as their resistance to sustainability is the result of justifiable fear of this discipline, the state does not so much encounter resistance at the hands of the market parties but as a result of the power of the market itself. The fisheries sector once again provides ample illustration of this. The overfishing of flatfish in the North Sea and the possible danger of exhausting these fish supplies can be described as a problem of overcapacity. There are too many boats in relation to their size and the level of technology available. This overcapacity is the result of a very costly race to obtain the best technology, a battle in which the fishermen became caught up after the Second World War. Not a single fisherman was forced to take part in this race at gunpoint. Yet at the same time the fishermen were not free in their choice: anyone who did not take part would be left behind technologically, thereby jeopardizing his chances of economic survival. The right technology and the relative chance of a good catch are closely related in the flatfish fisheries sector.

The hindrance that the state may experience in relation to the power of the market as such should be sharply distinguished from hindrance resulting from the will of particular businesses. This is especially true when the exercise of power is viewed from a public perspective. Where the rationality of the market is in play, we cannot expect too much from the isolated market actor. Under such circumstances she is just as helpless as any other actor in the face of a collective issue.

In order to be able to clearly distinguish between these forms of market power, I would like to introduce the concept of 'discretionary space'. The concept comes from the juridical theory regarding the public administration. It refers to the area where the public servant has freedom to act within the powers bestowed upon him. In keeping with Charles Lindblom (1977: 152-157), I wish to apply this concept to the domain of the market. Discretionary space indicates the sphere in which market parties enjoy freedom to act within the boundaries of market rationality, the law and standards of decency.

Legitimacy under pressure

Many theorists feel that the legitimacy of the state also suffered in the 20th century. Some commentators even speak of a crisis of legitimacy (Offe, 1979; Held, 1986; Bovens, 1996). This is not simply a way of saying that the state is losing some of its authority. The idea of a crisis of legitimacy stands for a downward spiral of declining success, decreasing authority, an even steeper decline in success, and so on. Chains of cause and effect are woven together here in a complex pattern.

The decline in the legitimacy of the state is the third main argument wielded against the indirect responsibility model by the administrative theorists. In Section 3.2 I divided the notion of legitimacy into four components: democracy, prudence, *Rechtsstaat* and effectiveness. I will unravel the tangle associated with the legitimacy crisis on the basis of these components.

Legitimacy under pressure: the Rechtsstaat
Within the context of liberal-democratic thinking, the idea of a *Rechtsstaat* boasts a great deal of authority. It stands for the victory over the executive power achieved by the ordinary citizens in the course of the 19th century. However, many a present-day author feels that there is also another side to the *Rechtsstaat*. It paralyses the workings of the state, thereby eating away at its credibility and effectiveness. One of the writers to make this point is Lindblom. According to Lindblom (1977: 346), the many checks and balances that exist are blocking the arteries of American administrative practice. Policy changes, especially those that require changes to the status quo, hardly ever get off the ground. Yeager's account of the endless series of

proposals, amendments, votes and debates leading up to the approval of the United States water purification act in 1972 illustrates Lindblom's point perfectly.

Legitimacy under pressure: democracy

It should come as little surprise that the ideal of democracy is coming under increasing pressure in today's society. Lincoln's concept of democracy as the pursuit of a government 'of, for and by the people' is difficult to realize under modern-day circumstances. Critics claim that practical circumstances block this ideal in three ways. In their eyes (a) the representative system makes a farce of the sovereignty of the people, (b) the complexity of modern administrative processes takes the meaning and purpose out of consultation and, as if this wasn't enough, (c) the hierarchy within the state has also disappeared. In this section I only wish to pursue this last criticism in greater depth, since this is the main criticism levelled by the administrative theorists. The other points of criticism will be dealt with in the following chapter.

In a representative democracy the political section of the state should lead the administrative section. As soon as the administrative section gains a certain level of autonomy, this immediately puts democracy at risk. In a representative democracy after all, only the political section of government is elected. In this regard, Mashaw (1985: 15 et seq.) speaks of the 'transmission belt' model. Many administrative theorists feel that reality is a long way from this ideal. The transmission belt is in fact turning in the opposite direction. The civil service has developed into a powerful organization which acts with far greater autonomy than the indirect responsibility model could ever allow. Policy is not prepared but cooked up in the administrative section of the state (Crince le Roy, 1971).

The civil service derives its power largely from the fact that it is quite simply impossible for parliament and the government to take all the decisions they ought to and to subject them to the necessary scrutiny. The quantity of these decisions alone means that the civil service must be allowed a considerable discretionary authority. In addition to this, dealing effectively with modern-day political and administrative issues demands a great deal of knowledge, experience and time. The administrative section has all of these elements at its disposal in far greater measure than the political section.

When attempting to explain the power of the civil service, many contemporary authors still appeal to the anti-bureaucratic literature of the 18th and 19th centuries. There the civil service is represented as a social tumour, with incompetent power-hungry civil servants as its cells (see Albrow, 1970: 17-26). These types of images are not very instructive, however, when it comes to explaining the problem of the discretion or the autonomy of the civil service. In my view, the problematic relationship between the political and administrative spheres mainly has to do with the transformation from a social order that is managed by control of limiting conditions to a social order that it forced to plan and deal with qualifiedly disfavoured conduct. As a result of this transformation the very character of government has changed. The exemplary situation of control by limiting conditions, in which those in political power issue a law or regulation which is then implemented by the civil service is becoming less and less appropriate. Government nowadays consists of interpreting a situation within the context of conflicting rules, values and facts. To govern is no longer to take decisions, as Kooiman claimed in 1971. Government is a process of continual action. Under these circumstances the discretionary space of the civil service inevitably expands.

Legitimacy under pressure: prudence
In this study the prudence of the state is defined as the extent to which it is capable of giving sufficient expression to the idealistic component of government. Many authors feel that the state has not been successful enough in this respect. Too much emphasis has been placed on the strategic component (Lindblom, 1977; Offe, 1979). The various authors are fairly unanimous as to the cause of this limited prudence. Partly as a result of its need for a process of give and take, the state cannot protect itself sufficiently against short-term interests and lobbies. This entanglement with narrow interests seriously erodes the prudence of the state.

Liberal-democratic literature frequently links the lack of balance between the strategic and the idealistic component of government with the theory about the relationship between the political and administrative sections of the state. The theory of Max Weber (1921) illustrates this point well. Weber operated on the assumption that the political section of the state would be particularly susceptible to narrow interests. In his eyes therefore it was not necessarily a bad thing for the citizens of a liberal democracy that the civil service had a measure of autonomy. As he saw it, a degree of autonomy for the civil service increased the chances of government based on sound judgement. Many authors nowadays invert this line of reasoning. According to authors like Paehlke (1989) and Cohen and Arato (1992) it is the civil service that is susceptible to narrow interests, especially those of market parties. These authors call for democracy and political control to be deployed in order to loosen the entanglement with narrow interests.

As I see it, both these views are inaccurate. The state cannot be divided into one part that is sensitive to narrow interests and another that is not. Both parts are sensitive to such influences in varying degrees at various times. A Dutch example which neatly illustrates this varying sensitivity is the so-called 'manure problem'. From the 1960s onwards, manure was applied to farmers' fields in such large quantities that it became one of the major causes of acid rain. The Ministry of Agriculture, Nature Management and Fisheries was aware of the negative effects of manure as early as the 1970s. Yet, for decades, the civil servants working for the ministry blindly protected farmers' interests and refused to take the relationship between manure and environmental problems seriously within the context of the public debate. Partly as a result of immense pressure from the Dutch parliament, the ministry eventually changed its tune in the 1980s. In the late 1980s and early 1990s, various plans for tackling the manure problem were formulated, including plans for the mandatory reduction of the number of pigs kept in the Netherlands. Egged on partly by pressure from the pig-farming lobby, parliament opposed the plan, using the argument that the impact of this measure on the pig-farming sector would be too great.

Legitimacy under pressure: effectiveness

The legitimacy of the state has also come under pressure in modern-day society as a result of problems relating to the administrative process. Many administrative analyses devote a great deal of attention to these management problems. I address them here under the heading of effectiveness. A first management problem the administrative theorists point to is the task of processing the immeasurably large mountains of information which confront the state. In this respect, Van Gunsteren and Ruyven (1995) describe modern-day society as 'the unknown society'. For administrators and politicians, this society is only accessible through a misty wood of global estimates, complicated graphs and tables where the key figures stand for the averages of averages.

A wonderful Dutch example of this phenomenon is the 'standard gauge project' (*'graadmeterproject'*) initiated by the Nature Management Department of the Ministry of Agriculture, Nature Management and Fisheries at the end of the 1990s. The intention behind this project was to summarize the Dutch natural environment and the policy related to it in a number of key figures that could be measured and compared on an annual basis. These key figures were aimed at giving citizens and politicians an insight into the state of the natural environment and into advances in nature management policy. In the development of these gauges, however, a dramatic trade-off took place between the aggregation level and the meaningfulness of the key figures. Nine themes were identified with around six gauges for each theme. The resulting figures were so general that they could only be used as a rough indication at best. Nonetheless, people were required to wade through 54 of these figures every year to gain any insight at all into nature policy.

The adequate processing of information is not the only management problem facing the state. Disagreements over jurisdiction, and cultural and political differences of opinion between ministries also take their toll. Administrative theorists refer to 'sectoralization' or 'the fragmented state' (Andeweg, 1985; Edelman-Bos, 1986). Meanwhile the highly collective dimension of activities and processes means that unintended consequences spring up all over the place. Even something as idyllic as wind energy has negative effects. If their location is not well-chosen, wind turbines can ruin the landscape and result in the wholesale slaughter of birds.

Inappropriate powers: micro planning

The last main argument that administrative theorists use against the feasibility of the indirect responsibility model is that the powers of the state are not suited to the kind of control required in the modern context. While it is true that the state has mighty means at its disposal to impose its will, this power does not reflect the nature of the management requirement. The instruments wielded by the state are no longer appropriate. Reconstructing the arguments of the administrative scholars, two main causes of this incongruence or imbalance between administrative potential and administrative requirements surface (Kooiman, 1988). One argument centres on the need for micro planning, the other on the loss of the vitality of the law. Both arguments will be dealt with briefly below.

In order to explain the argument relating to micro planning, I would first like to introduce the concept of 'ecological modernization'. This concept is often defined from an environmental perspective. For example, Tuur Mol and Gert Spaargaren (1991: 18) state that we should understand ecological modernization as the ecological reconstruction of production and consumption processes. However, I would like to approach this concept from an institutional perspective. In that case we can describe 'ecological modernization' as adapting the production processes in such a way that more account is taken of the collective dimension of activities and processes. When described in this way, it immediately becomes clear why the powers and the might that the indirect responsibility model accords to the state do not fit in with the kind of management sustainable development demands. Ecological modernization requires that considerable attention be paid to public considerations within businesses; on the workfloor, in the laboratories and at the board of directors' table. In short, ecological modernization calls for planning at meso level and micro level. Only then can a balance be found between economic efficiency and taking care of the collective dimension of activities. However, liberal democracy and the indirect responsibility model are opposed to this type of government. The powers needed to manage processes at this level are therefore extremely limited and moreover hindered by principles like the division between public and private domain and the demand that government activities be lawful.

One of the authors who writes admirably about this inappropriateness of the instruments of government is Charles Lindblom. According to Lindblom (1984) many decisions relevant to government are taken in the boardrooms, marketing agencies and research laboratories of big businesses, due to the unintended consequences of corporate activities and processes. It is there, says Lindblom, that the real policy makers are to be found. After all, that is where micro planning takes place and where it is determined what kind of products will be made, which technologies will be invested in and so on.

Lindblom's claims can be clarified by our case study of the Dutch flatfish fisheries sector. Once it had dawned on the Ministry of Agriculture, Nature Management and Fisheries at the beginning of the 1980s that it was necessary to limit flatfish fishing in the North Sea, a large number of measures were taken within a short space of time. The number of ships and the maximum engine capacity were regulated, rules were introduced for the length of the fishing vessel (the beamed trawler), and the number of days that a ship could put out to sea. None of these measures was sufficient. The ink on the last decree was barely dry before the need for a new regulation became apparent.

It should come as no surprise that in this case the state did not succeed in gaining a hold over the 'private policy makers'. Fishermen responded to regulations. A new law or a new regulation was only regarded as a further reduction of the area within which they were free to search for the most economically efficient solution. If the state attached limits to the capacity of a trawler's engine, the fishermen would go in search of lighter craft in an attempt to maintain their fishing capacity. If the size of the mesh was restricted, the fishermen would take to fishing with double nets, and so on. The fact that, at the end of the 1980s, as legislation continued to pile up, the state still had trouble controlling the sector illustrates the problem that Lindblom identified. Solving it calls for planning at the micro level. However, in a liberal democracy the state is not allowed to intervene at this level. And in the 1980s, the fishermen themselves did not take public considerations into account when it came to technological innovation, investment decisions and day-to-day practice.

The stalemate in relation to micro planning is something that often occurs in present-day markets. There are plenty of such examples around, and no wonder. In a market where the state is only allowed to determine limiting conditions and where the responsibility of businesses is limited, there are great incentives for companies to act as rationally as possible in economic terms. This economic rationality does not take collective, unintended consequences into account. The chance that the market will spontaneously set out on a path of ecological modernization under such conditions is therefore slight.

Inappropriate powers: the law

The imbalance between the potential for government control and the need for government control should also be linked to the fact that the state in today's society comes up against the limits of the law. Administrative theorists argue that the law is losing its power as an instrument of administration. It often just doesn't seem to work anymore. It operates as a screwdriver where a hammer is needed. For the indirect responsibility model in particular, this is a worrying development. Within this mental model of social organization, the law is the obvious instrument of control (Hafkamp and Molenkamp, 1990: 212). I will discuss these legislative problems under the heading 'the law's loss of vitality'.

The law's loss of vitality is first related to corporatization or the rise of the corporate society. By corporatization I mean the process by which organizations increasingly become relevant actors in society (Dewey, 1905: 443). The law is poorly equipped to deal with these non-human legal persons. Organizations respond to the law differently than natural persons. Christopher Stone (1977) formulates the cause of this clearly and concisely: we cannot put organizations in jail or sentence them to death. More specifically: a typical feature of the organization as actor is that it can only act through its representatives. This makes it difficult for the law to grapple with an organization. It becomes difficult to allocate responsibility in a meaningful way. An organization as such has 'no body to kick and no soul to damn'. It cannot be a moral actor. Meanwhile, it is often very hard to translate deeds done in the name of an organization into the responsibilities of the natural persons who operate as its representatives. In organizational theory this is known as 'the problem of many hands' (Thompson, 1980). At the heart of this problem lies the phenomenon that responsibility within an organization easily 'evaporates'. There are always several people involved in any given action. This makes it difficult to establish clear causal relations between the actions of the organization and those of specific natural persons. In addition to this, it is also often tricky to establish moral relations. People within organizations are frequently under a measure of duress and are often not fully informed about processes that are going on in their organization.

An entirely different consideration which makes it difficult to call companies to order, is the fact that a company drags many people down with it when it falls, including employees, residents of economically deprived areas and so on. These innocent parties as it were protect the business from being given the hardline treatment.

The law's loss of vitality can also not be seen separately from the technologization of society (Bijker and Law, 1992; Law, 1991 and Hughes, 1983). By technologization I refer to the process by which technology and technological developments are having an increasing influence *'on the way in which we live and organize society'* (Bijker: 1992, 2). An important result of technologization is that people's actions are becoming more and more dependent upon the technologies at their disposal and of which they make use. Technologization therefore implies that the results of social processes are less dependent upon human behaviour. Controlling the market therefore means controlling the effects of technology. The law was not designed to cope with such a task. It is primarily geared towards restricting and steering human behaviour.

A further reason for the loss of vitality of the law is closely related to the above observation. The approach to modern public problems often demands a considerable degree of coordination within a business column or between various sectors of society. The law is not designed with this coordination in mind. The management of the water levels in the Dutch nature reserve De Grote Peel is a good example (Glasbergen, 1989). De Grote Peel is a moorland area which requires a high level of groundwater. This can only be maintained if the surrounding areas also maintain a relatively high groundwater level, as groundwater has the annoying characteristic of spreading itself around in equal measures. However, the farmers in the area benefit from a low groundwater level, as it stops their tractors sinking into the mud during the wet spring season. A conflict of interests like this is difficult to solve by the drawing up of regulations alone. It requires that all the administrative organizations concerned harmonize their policies with each other and with the farmers.

Developments such as the 'proceduralization of law' and the constitution of legislation in 'framework laws' have also put the vitality of the law severely to the test. Both processes can be seen as the law reacting to the increasing complexity, diversity and dynamism of modern society. Authors like Ingeborg Maus (1986) and Jerry Mashaw (1985) are worth mentioning in this context. The former writes about what is often called the rise of framework law, the latter about the increasing proceduralization of the law.

Maus positions her argument within the context of Jürgen Habermas's thesis that society is becoming more juridified and that there are normative objections to this process. Maus augments this argument by proposing that at the same time a process of *de*juridification is taking place. This process is due to the fact that laws contain less and less clearly defined norms for conduct. Laws are turning into so-called framework laws: laws which only become meaningful by virtue of an additional layer of content drawn up by the civil service. The Dutch legislation on animal health and welfare (Handelingen der Tweede Kamer, 1981, 1985 and 1986) is a good example of this. It prohibits the keeping of animals, except for those included in a special list drawn up by the executive power. The list includes almost all animal species that occur naturally in the Netherlands.

According to Maus, the process of dejuridification leads to a situation in which the law loses its teeth. Laws no longer have clearly defined implications. An impressive-sounding law against environmental pollution no longer implies that the environment will actually become cleaner. Another consequence of all this is that it becomes more difficult to check the conduct of the state. Maus sees this as a danger to the legal certainty of businesses. And it is not only the legal certainty of businesses which is at stake. It also becomes more difficult for non-governmental organizations to use the law to force the state to protect the interests of the actors or lifeforms it ought to be protecting.

Mashaw does not speak of dejuridification but of an ever increasing proceduralization of the law. However, when it comes to content, the authors are actually very closely related in their arguments. Mashaw posits that the proceduralization of the law leads to a situation in which all behavioural norms come adrift.

Albert Hirschman (1992) points to one last cause of the law's loss of vitality. According to Hirschman we nowadays have a strong tendency to ignore the fact that in a liberal-democratic structure, the main function of the law is symbolic. The instrument of the law only works effectively if it can be a normative and judicial compass for an independently operating actor. If the law is confronted with consistently unwilling actors who constantly have to be forced to conform, then it is dead on the page. Enforcement costs will then hit the roof and continue to rise. This will be especially true of victimless crimes. An everyday example is the Dutch regulation that dogs should be kept on a leash in nature reserves. In actual fact, dogs are usually the only animals one is likely to see running around free in a Dutch nature reserve on a Sunday.

In the realization that administrative law has developed into a vast, detailed and technical discipline, Hirschman's point is highly relevant. What seems an entirely logical development is actually damaging the symbolic function of the law. Technical rules, such as the European regulation that fish-processing companies have to have tiles on the wall up to a certain height and that the tiles have to be a certain size, thickness, colour and quality have a less extensive symbolic or normative function than general rules which follow on from our everyday sense of morality, such as the law forbidding murder. Laws which spiral out of control for technical reasons can easily go off the rails completely and lose credibility, symbolically speaking. They become part of an abstract reality, an ever increasing mountain of paper under the direction of an abstract regulatory bureaucracy that sometimes comes up with rules which are nonsensical when applied to real-life situations or with which it is impossible to comply.

Implication: beyond the indirect responsibility model

To sum up: administrative theorists heavily criticize the indirect responsibility model. Basically their arguments come down to the fact that the indirect responsibility model is based op a number of invalid assumptions. Four of these are:
- the state is no more an *all use institution* than the market or civil society. It can most definitely be overloaded;
- the state is not so independent from the other fundamental institutions that it need pay them no heed when exercising its powers;
- the state does not have an endless source of legitimacy at its disposal;
-   the state does not have powers tailored to suit its situation.

Taking these criticisms as a whole, the administrative critics feel there is only one conclusion to be drawn. The indirect responsibility model no longer fits within the context of modern society. Following on from this, they go on to conclude that the institutional order of modern society, which is structured according to the indirect responsibility model, is also in dire need of revision.

3.5 AN ADMINISTRATIVE THERAPY

According to administrative theorists, liberal democracy is in need of a new mental model of social organization. The indirect responsibility model has had its day. In administrative theory we encounter various new designs for such a new mental model. In this section I turn my attention to just one these. Following in the footsteps of Pinkerton (1989) I have named it 'co-management'. At the heart of co-management lies the notion that greater use should be made of the market parties themselves when it comes to managing the collective dimension of action. In this 'age of unintended consequences' (Ankersmit, 1995), the market agents can no longer be spared from a public point of view. Market parties should take on a certain responsibility for public issues.

Within modern-day administrative theory concerned with sustainability and market control, co-management is, in my opinion, the cream of the crop. First of all, this is because advocates of co-management do not reduce the problems of modern government to internal problems of the state. In the eyes of the advocates of co-management, 'ungovernability' is a problem of the institutional order and does not come about as a result of inefficiency or internal coordination problems within the state as such. Another important consideration is that the advocates of co-management do not place the desirability of societal steering itself on trial, as some other administrative theorists tend to do. Advocates of co-management are imbued with the notion that the necessity of societal steering is directly linked to the liberal-democratic pursuit of a *well-organized society* (Glasbergen, 2001).

Co-management

Co-management currently enjoys great popularity within the academic discipline of public administration. In the Netherlands, we can find this school of thought represented by such environmental administrative theorists as Wim Hafkamp and Molenkamp (1990), Pieter Glasbergen (1989b, 1992, 1995, 2000), Jan Kooiman (1993a, 1993b and 1993c) and Martijn van Vliet (1992 and 1994). Yet co-management is more than just a Dutch affair. It has also been welcomed with open arms on the international scene, by the likes of Britain's Albert Weale (1992), Germany's Martin Jänicke (1997), Norway's Sven Jentoft (1989, 1995, 1998, 2001) and Canada's Emily Pinkerton (1989). However, it is interesting to note that a number of these international authors point to the Netherlands as a guiding light in this respect. Jänicke for example describes the Dutch attempt to involve market parties in political-administrative processes as an administrative innovation whose example deserves to be followed.

While I will go on to speak of *the* advocates of co-management and *the* theory of co-management, the reader should bear in mind that co-management is not something that has crystallized into a single coherent design. Thinking on co-management still finds itself very much in the early stages. Various authors give their own interpretation of this theory in the making. We cannot therefore speak of a single coherent argument for co-management. One way in which this variety manifests itself is in the range of names given to this concept. The ideas surrounding co-management are presented under different titles. Examples include 'communicative control' (*'communicatieve sturing'*, Van Vliet, 1992), 'usergroup participation' (Jentoft, 1995), and 'social learning' (Glasbergen, 1997).

The first fundamental innovation that the advocates of co-management introduce into the indirect responsibility model is the utilization of the market actors themselves for the handling of public issues. They depart from the notion of the market as a sphere of limited public responsibility. Instead market actors are allocated tasks aimed at making the market more sustainable. These tasks relate both to the process of policy preparation and the processes of implementation and enforcement.

The second fundamental innovation in this regard concerns the link between the market and the state. Where the advocates of the indirect responsibility model draw a sharp line between these two fundamental institutions, the advocates of co-management envisage a system in which market parties and the state maintain intensive contact with each other, exchange information, discuss and cooperate. Advocates of co-management are not hostile to the idea of setting up institutional structures with this very objective in mind. In their eyes, a close-knit interweaving of intermediary organizations is indispensable for achieving adequate control over the market within the modern context.

These important innovations are closely related. According to the advocates of co-management there is no hope of market parties taking independent action to initiate a process of ecological modernization within the existing institutional system. It is necessary for the state to initiate, stimulate, inform and compel. In addition to this, the state is needed to formalize the arrangements made, legal and otherwise.

This recognition of the lasting significance of the state makes the difference between co-management and naive interpretations of self-regulation. In its more naive versions, self-regulation is a mental model that stands for a system in which companies attempt to function as responsibly as possible, independently of one another and the state. The advocates of co-management recognize that this option is beset by many limitations when it comes to finding solutions to public issues. If self-regulation is to become a success with regard to complex issues such as sustainability, some degree of coordination and cooperation between market parties is inevitable. The state therefore has a role to play.

Thinking on co-management shows quite a few similarities to the theory of corporatism. At the centre of the corporatist theory, after all, there is also a notion that market parties ought to play a certain public role and that mutual cooperation between market parties and their cooperation with the state are of great importance. In Wyn Grant's (1985: 3-4) definition of corporatism, this family resemblance is clearly visible.

> '*I use the term "corporatism" (as a shorthand for liberal or neo-corporatism) to refer to a process of interest intermediation which involves the negotiation of policy between state agencies and interest organizations arising from the division of labour in society, where the policy agreements are implemented through the collaboration of the interest organizations and their willingness and ability to secure the compliance of their members. The elements of negotiation and implementation are both essential to my understanding of corporatism.*'

Despite this obvious kinship it would be inaccurate simply to reduce co-management to a (neo-)corporatist variant. Differences are also evident. Corporatist thinking is strongly allied to social-economic issues. Advocates of co-management tackle other public problems, seeking out other social coalitions and alliances in the process. The tripartite consultation between the state, market parties and employees is replaced, for example by a consultation between state, market parties and representatives of environmental groups (Weale, 1992). Advocates of co-management also place less emphasis on the formalization of structures within the market. Cooperative structures may be fluid.

It would be interesting to know how the advocates of co-management themselves view their relationship with corporatism. However, they are often silent about this family relationship. Only Jänicke (1997: 73, 76) sets out an explicit connection between neo-corporatism and his own approach. He has no objections to labelling his approach as a form of neo-corporatism.

Advantages and implications of co-management

At first glance the notion of giving market actors a part to play in reducing the negative public effects of the market process doesn't seem like a bad idea. Many control-related problems would it seems be alleviated by getting market parties directly involved in the political-administrative process. To start with, the burden on the state becomes lighter with every task that market parties can take over. What is more, the state's dependency on the market is of less significance if the fundamental institutions work together. An aspect that the advocates of co-management themselves continually emphasize is that contributing to the process of thinking about and deciding on environmental policy will increase the loyalty and commitment of market parties with regard to public policy (Jentoft, 1989: 139). The judiciousness of environmental policy could also be increased by co-management, at least to the extent that the lack of it has to do with gaps in the knowledge and information available to the state. An environmental policy drawn up in cooperation with market parties will, in all probability, produce a better balance between economy and sustainability than a policy put together by civil servants who are positioned outside of the actual circumstances to which the policy applies. There is another advantage in a similar vein: the transfer of tasks to actors within the market will probably facilitate in a more integrated approach to environmental problems (Hafkamp and Molenkamp, 1990). The most important advantages of co-management, however, appear to be its ability to break the stalemate with regard to micro planning and to compensate for the problems relating to the law's loss of vitality. After all, when market parties are involved in solving public problems, then public considerations have entered the workfloor, the boardrooms and the research laboratories of the business community, internalizing sustainability as part of the economic process.

Differences between co-management and the indirect responsibility model.

But on what points exactly does co-management differ from the indirect responsibility model as a mental model of social organization? While the literature on co-management contains no systematic description of these differences, it is nonetheless possible to draw up a list. The most important difference between the indirect responsibility model and co-management is that co-management does away with the idea that actors on the market only have a very limited public responsibility for the collective dimension of their actions. Companies are expected to take on the role of publicly responsible parties. They are no longer permitted to concern themselves only with their own business interests. They should be prepared to take action which is not entirely rational from a purely market-oriented perspective. In this respect, Van Vliet (1992: 69) speaks of the company as a 'reasonable citizen' (*'redelijke burger'*), while Hafkamp and Molenkamp (1990: 240) call the company a 'responsible citizen'.

Secondly, co-management entails a transformation in the way in which state and market interact. In this connection, co-management embraces 'communicative management'. Van Vliet (1992: 24) described communicative management as a '[policy] *process in which the negotiations (between opposing parties) are adequately structured'* with the explicit goal of letting *'"sound argument"' be the decisive factor as much as is possible'.* Analytically, communicative management can be divided into two dimensions. Firstly, it is a vision of the way in which laws and policy ought to be *implemented.* As such, the concepts of 'voluntariness', 'appeal to morality', 'tailor-made approach' and 'mutual agreement' are central to the approach. But communicative management is also a vision of the way in which laws and policy ought to be prepared. Central to this dimension of communicative management are concepts like 'cooperation', 'consultation', 'dialogue' and 'compromise'.

Both components of communicative management clash with the indirect responsibility model. Where the phase of policy preparation is concerned, the indirect responsibility model calls for a distance between the state and market parties. After all, social differentiation means that the state has to concentrate on its own rationality. Pressure from market parties can only detract from the state's own rationality. Co-management rolls up and puts away this stark division of labour between the market and the state.

Where the implementation of policy is concerned, the indirect responsibility model stands for control by limiting conditions. Control by limiting conditions means that the state only draws up general laws, which it upholds by the threat of force. A typical feature of such laws is that they do not prescribe positive behaviour, but set a bottom line for acceptable behaviour. They define what is not allowed. Within the space set out by the limiting conditions, a market actor is free. Co-management creates a whole different ideal. It aspires to a situation in which market parties, together with the state, actively search for the best course of action in the light of both economic and public considerations.

The sort of legislation which would fit in well with co-management is described in detail by Teubner and Willke (1984). According to Teubner and Willke the law has to adapt to the fact that companies have become relatively independent and self-referential systems.

> '*This form of control is the first to draw the necessary conclusions from the phenomenon of self-reference and the circular closure of complex living systems which is by no means new but has received a new theoretical basis in the theory of autopoietic systems. Any attempt at purposive intervention in such systems – be it in the form of therapy, planning, control, development policy or the like – must presuppose this basic self-reference, work with it and include it as part of one's own generation of action.*'
> (Teubner and Willke, 1984: 33)

Teubner and Willke refer to the type of law that fits the context of modern society as 'reflexive law'. The term refers to a procedural law which encourages or obliges actors to evaluate and/or amend their own conduct from a reflexive point of view. In this regard, reflexiveness or a reflexive orientation is defined as:

> '*... an individual or collective actor's capacity for empathy ... i.e. the ability to put yourself in the role of other actors in order to see your own role from their point of view. In its generalised form, reflexiveness is a form of self-control which allows sub-systems to thematise their own identity and adjust precisely to meet the requirement that they have to be a suitable environment for other subsystems that act in relationships of interdependence in the environment relevant to the actor.*' (Teubner and Willke, 1984: 23; the authors cite Luhmann here.)

A law that takes advantage of such a reflexive capacity is a procedural law.

> '*Reflexive law tends rather to more abstract procedural programmes which concentrate on the meta-level of the regulation of processes and organisational structures, on the assignment and redefinition of rights to control and on decision-making competences.*'
> (Teubner and Willke, 1984: 23)

It is quite possible to imagine that legislation of this kind fits in well with the need for micro planning. However, it does not rhyme at all with the notion of control by limiting conditions. Due to the emphasis on voluntary concern with publicly desirable action, it can even be seen as the opposite to some extent.

For the sake of accuracy it should be noted here that the advocates of co-management do not completely dismiss the need for force, strict rules and tough measures. Glasbergen (1997: 185) regards it as essential that definite agreements be laid down in covenants. Hafkamp and Molenkamp (1990) too are of the opinion that firm and clear rules are vital. However, advocates of co-management do feel that regulations should be less detailed. They should be more geared towards defining targets that companies are required to reach in the long term and towards setting out procedures to be followed.

Thirdly, co-management and the indirect responsibility model differ in their views on the state. Just as market parties are expected to show far-reaching involvement in public considerations within co-management, so the state is required to show more empathy for and involvement in the economic situation in which the companies find themselves. In addition co-management requires greater flexibility of the state when it comes to applying regulations and the like. Of course, the indirect responsibility model also dismisses this form of dedifferention.

A fourth and surprising difference is that co-management bestows a role upon civil society in the control of the market. The neglect of civil society, so typical of the indirect responsibility model, is broken by the advocates of co-management. As the co-managers see it, the relevant parties from civil society should take up a place at the table where the market parties and the state hold their consultations about the policy to be adhered to (Weale, 1992: 167-180). In addition to this, the institutional possibilities available to civil society in order to get companies to change their behaviour should be expanded. These might include changes to liability laws or to regulations which demand greater openness in business affairs (Weale, 1992: 177).

These suggestions can be interpreted as attempts to make civil society publicly, politically and administratively relevant. The protagonists of co-management do not go very deeply into the reasons for this remarkable rehabilitation of civil society. Yet it is possible to distil a number of these reasons from the literature. Involvement on the part of civil society increases the prudence and the learning capacity of government (Jänicke, 1997: 79). In addition to this, the advocates of co-management are aware that their proposed changes to the indirect responsibility model contain the danger of a democratic shortfall. The involvement of market parties in public processes also presents the danger of companies gaining an increasing say in the way things should be run. This spreading influence can come into conflict with the pursuit of democracy. The mobilization of civil society for the handling of public issues is a manoeuvre aimed at avoiding this clash and/or somehow softening the blow (Glasbergen, 1997: 190). In addition, advocates of co-management see involvement on the part of civil society as offering a sizeable degree of support for the state's goal of ecological modernization of the market (Hafkamp and Molenkamp, 1990: 238). Market parties will be easier to convince of the need for such developments if it is clear that the state enjoys the support of the citizens. The citizens' involvement in the public process therefore strengthens the state's negotiating position. An active civil society lends the state its authority, as it were.

Giving civil society a role in controlling the market is also related to what may be called the fragility of liberal democracy. The advocates of the indirect responsibility model talk about liberal-democratic values as if they can be taken for granted. Advocates of co-management, however, realize that these values always have to be experienced, learnt and developed anew, and they feel that civil society is the best environment for this.

Co-management: a moderate proposal?

Co-management has become known as a moderate set of ideas: a therapy which tries to take what is politically and administratively feasible as its starting point and which, for the sake of feasibility, is prepared to make concessions regarding what is desirable. This moderate image is confirmed by the fact that co-management or similar ideas that go under different banners have had a great deal of practical influence on politics in the Netherlands, especially when it comes to sustainability. The Ministry of Agriculture, Nature Management and Fisheries and the Ministry of Housing, Spatial Planning and the Environment, two key ministries when it comes to sustainability, have adopted many of the ideas relating to co-management. Pieter Winsemius, who was Minister of Housing, Spatial Planning and the Environment in the 1980s even wrote a book on politics and administration during his time in office, which fits perfectly within thinking on co-management. For a long time, the Dutch environmental movement, and its more radical sections in particular, were therefore somewhat sceptical about this alternative to the indirect responsibility model. In their view the looming environmental crisis could not be dealt with effectively within such a moderate political-administrative approach.

I think that the estimation of co-management as a moderate set of ideas is partly wrong. In order to get a clear view, however, we have to make a distinction between co-management as a practice and co-management as a mental model of social organization. As a practice co-management may be called a moderate therapy, certainly in the Dutch corporatist context. Many of the ideas propagated by the advocates of co-management go back a long way in Dutch governmental practice. As a mental model however, co-management is most definitely something radically new. As we have seen, it constitutes a frontal attack on some of the dogmas of the indirect responsibility model.

3.6 AN EVALUATION OF ADMINISTRATIVE THEORY

Administrative theory is of great value to the aims of this study. It criticizes the indirect responsibility model explicitly and it outlines the contours of an alternative mental model. All the more reason then to subject it to a critical evaluation in this section. For starters, it should be noted that the administrative theorists have an impressive array of winning cards to play in their criticism of the indirect responsibility model. However, as I see it, they fail to fully capitalize on their advantages. Quite a few administrative arguments are not rigorous enough from an analytical point of view. Many administrative theorists claim they want to show the inadequacy of *thinking* on government control, while in their analyses they support these arguments by criticizing modern-day administrative *practice*. However, someone setting out to show that sustainability calls for a new mental model of social organization has to observe a careful separation of these two levels of analysis. After all, a moderately or poorly functioning administrative practice does not constitute sufficient proof of the inadequacy of a mental model. In order to criticize this effectively, it is necessary to show that the mental model suffers from inherent problems (see Section 1.2).

The confusion of theory with practice means that this awareness is missing and this represents a shortcoming. Advocates of the indirect responsibility model are not necessarily shocked by the cartloads of empirical case studies produced by the administrative theorists to show that modern liberal democracy is suffering from serious administrative problems. They do not claim a perfect fit between mental model and practice. What characterizes an advocate of the indirect responsibility model is the conviction that, when practical problems occur, the indirect responsibility model can point the way towards improving the institutional order: he believes in the regenerative power of the indirect responsibility model. Herman Daly, who has already been presented as just such an advocate, wholeheartedly endorses much of the empirical criticism from administrative quarters. In *For the Common Good* (1989) he shows that the state does not determine the limiting conditions of the market process independently, that it negotiates with market parties, that environmental laws are often lacking, that their enforcement is often woefully inadequate and so on. Yet for Daly, all these discrepancies between the model and political-administrative practice are no reason to abandon the indirect responsibility model. As he sees it, the market can and should be made more sustainable by restoring the indirect responsibility model. Political and administrative practice should be made to function in accordance with the indirect responsibility model as it was meant to be. The state must once again firmly set the limiting conditions of the market process. If administrative theorists are to succeed in convincing Herman Daly that he is mistaken and that society is in need of a new mental model, then they will have to come up with an additional argument. They will have to make it clear that the modern context has eaten away at the viability of the indirect responsibility model.

We can also express reservations about the administrative theorists' critique of the indirect responsibility model from a historical perspective. Again, the confusion of theory with practice hinders the diagnosis of the administrative theorists. Administrative theorists often state that today's society is struggling to cope because we only have 19th century methods at our disposal. If we look at the actual course of events, however, this statement is obviously incorrect. Western societies have responded to the threat of administrative decline in all sorts of ways. For example all kinds of attempts have been made to adapt the juridical armamentarium to the demands of the times. Financial sanctions and the development of framework legislation are just two examples of this response. We might even go a step further and assert that the turbulent development of the field of administrative law in the 20th century has been largely bound up with efforts to adapt the indirect responsibility model to the needs of the age. At the practical level, the argument of the administrative theorists is therefore untenable. However, at the level of theory, their position can be defended. The indirect responsibility model stems from the first half of the 20th century and does indeed rest on assumptions which are not very realistic in the present-day context.

A comparable historical weakness is also present in the administrative analysis of the link between market and state. Many an author gives the impression that 'once upon a time' in Western society this link was formed purely in accordance with the indirect responsibility model (see for example Bovens, 1996c). This statement contradicts the historical observation that the Dutch system, for example, has always had a strong corporatist element with much interweaving of state and market (Daalder, 1985). Nevertheless, the claim that the Dutch system in the 20th century can be described as hierarchic and bureaucratic need not be dismissed completely. If we observe the distinction between theory and practice, we can state that in practice corporatism was all around but that at the same time, at the level of mental models, the indirect responsibility model was the dominant view. Philippe Schmitter (1982: 275) provides arguments to support this layered view. Schmitter shows that corporatism in Europe has made its presence felt far and wide, but that at the same time a legitimation of this ideology is nowhere to be found.

Some historical weaknesses of the administrative diagnosis of the indirect responsibility model have nothing to do with the confusion between theory and practice. One particularly irritating shortcoming is that administrative theorists often fail to place their critique in its proper time frame. Are we talking about a process which has been going on for more than a century, as ideas relating to the rise of technology and corporatization suggest? Or, in line with the emphasis placed on the vicissitudes surrounding the enforcement of regulations, should we be thinking more in terms of the last 50 years?

A historical weakness worthy of a little more attention in my view is what I regard as the exaggerated nature of many of the administrative analyses. Reading through the administrative literature one sometimes has the impression that government in modern society is utterly useless and that the state just stumbles along, piling mistake on mistake. Seen empirically, such a view cannot be maintained. There are many areas in which the state does not perform badly at all (Goodsell, 1985). Think for example of education, infrastructure, law enforcement and social issues. It is perhaps going a bit far to sing the state's praises, but in many Western societies, its performance in these areas can certainly be described as satisfactory. In its pursuit of sustainability too, the state has booked considerable successes in recent decades. In many cases it did so within the confines of the indirect responsibility model.

> *'Trends were not all negative, of course. ... (In OECD countries) sulphur dioxide emissions fell by 25 percent between 1970 and the end of the 1980s. More importantly, perhaps, the link between sulphur dioxide emissions and GDP has been broken, so that trends in sulphur dioxide emissions in Canada, the US, Japan, France, Germany, Italy and the United Kingdom are now negatively correlated with trends in GDP. There has*

also been a marked decline in OECD countries in the concentrations of particulates in the atmosphere in urban centres.' (Weale, 1989: 25)

Bardach and Kagan express themselves in similar terms.

> *'Tougher enforcement of protective regulation has brought some tangible results. For example: ...*
> *   *Neither air nor water pollution control made major strides until the strict deadlines, permit systems, and enforcement strategies of the Clean Air Act of 1970 and until the Federal Water Pollution Control Act of 1972 transformed cooperation-oriented state agencies into enforcement-minded ones. By 1977, according to the Council on Environmental Quality, over 90 percent of the major water pollution sources in the nation had installed the "best practical technology".*
> *   *... As a result of aggressive regulation by the EPA and the FDA, the government announced in 1979 that traceable amounts of polychlorinated biphenyls (PCBs) in food, although probably impossible to eliminate entirely for some time to come, had dropped very considerably and were virtually absent in poultry and dairy products.'* (Bardach and Kagan, 1982: 93-95)

In the Netherlands too, success has been achieved regarding emissions of sulphur dioxide: total emissions between 1980 and 1997 were down by 40 per cent (Van der Straaten, 1990: 173). In addition to this achievement, Dutch victories in the battle against the largely organic pollution of surface water achieved national and international renown.

> *'Up to 1970, the Netherlands had achieved very little in the way of water pollution control. Situated in the Rhine delta, the Dutch were reluctant to spend money in the light of effects of transboundary pollution. In 1969, however, after many years of preparation, the Dutch parliament approved the Surface Waters Pollution Act, and the Netherlands soon emerged with Western Europe's most prominent and comprehensive scheme of water pollution control. During the first decade, the Netherlands had not only caught up with its most advanced neighbours, but also experienced a considerable reduction in industrial emissions.'* (Andersen, 1994: 146)

How much of the administrative theorists' diagnosis remains intact after all this criticism has been heaped upon it? As far as I can see, quite a lot of it emerges unscathed. To start with, we cannot ignore all the valuable empirical material collected by the administrative theorists, revealing that government control of the market is reaching its limits. This material is also supported by research and analyses from adjacent academic disciplines such as legal sociology and political philosophy. Turning to the administrative theorists' historical flaws, it must be said that the criticisms from that angle constitute more of a request for specification than a frontal attack. Even the analytical weak spots in their theory seem reparable. The most important point, i.e. showing that modern-day problems of government control have an interpretative dimension, can be tackled by making a consistent effort to separate theory from practice and by showing that the existing mental model of social organization is no longer sufficient. In fact, these repairs have already been carried through in the previous sections of this study.

A closer look at the therapy

Now that the administrative criticism of the indirect responsibility model has survived evaluation, it becomes all the more pressing to discuss the alternative put forward by the administrative theorists. Can co-management be presented as a fully fledged alternative to the indirect responsibility model? Here too it is important to maintain a clear separation between theory and practice. The question of whether the administrative theorists' ideas about co-management have been successful in administrative practice is quite different from the question of whether administrative theorists are able to formulate a mental model which serves as a fully fledged alternative to the indirect responsibility model. Yet in this regard too, administrative theory is found wanting in terms of analytical rigor. The often asserted claim that co-management has been worked out as a mental model is not substantiated, as co-management is chiefly presented as a practical strategy which enables the state to deal more effectively with public issues. This being the case, I will mainly evaluate co-management as a practical strategy in this section, before going on to transpose the conclusions of this evaluation to the level of mental models in the next.

The significance of co-management as a practical strategy can itself be interpreted in two ways. On the one hand, the degree to which co-management is actually embraced by administrators, public officials and others can be taken as a measure of its success. On the other hand, co-management's performance at a practical level can be assessed theoretically in terms of its systematic potential. With regard to the first interpretation, there can be no doubt as to co-management's success in some countries, the Netherlands among them. The idea of a shared public responsibility has been made one of the pillars of Dutch environmental policy. Historically, the success of co-management is easy to explain, since the Netherlands has long occupied a unique position when it comes to the mental model of social organization, a position in which liberal-democratic notions are effortlessly linked to a corporatist social design. In such a context, it is not hard to understand why the arguments in favour of co-management are so well received. However, this makes the independent effect of the plea for co-management itself difficult to measure. The Dutch context even gives rise to the question of whether it wasn't the administrative practice that formed the inspiration for co-management instead of the other way round. An additional factor here is that the plea for cooperation and shared responsibility in the Netherlands is not exclusively the province of co-management. Many other academics and administrative theorists preach similar ideas under a different flag. Whatever the exact ins and outs of the situation, it is true to say that present-day administrative thinking about co-management harmonizes well with developments in administrative practice. In this sense co-management is a success in practical terms. (For an English-language account of Dutch environment policy see Weale, 1992.)

This brings us to the systematic potential of co-management. In my view there is reason to be cautious here. In its present form, the design of the advocates of co-management is seriously flawed. One noticeable omission in the co-managers' design is the recognition of what Lindblom calls *'the privileged position of business'*. There is also little emphasis on the imbalance between the idealistic and strategic components of government. There are many other potential points of criticism as well. However, it is not my intention to give an extensive or comprehensive account of them here. Thinking on co-management is not yet fully developed and therefore a commentary which focuses on details and splits hairs would be out of place. I will therefore raise only one important point about its feasibility and permit myself a limited commentary from a liberal-democratic viewpoint. I see this normative appraisal of co-management as being more relevant to this study.

Discretionary room

Co-management takes it for granted that actors in the market have a certain freedom to act. After all, it is only sensible to demand that market actors take some responsibility for public issues, if one thinks that they are actually at liberty to do so. Many market theorists have grave doubts as regards this assumption. According to Hayek, in a well-organized market, the discretionary room for manoeuvre is heavily restricted by '*the hard discipline of the market*' (Hayek, 1980: 24). Focusing on 'self-regulation' - which in this respect is a lot like co-management - Ian Maitland too has brought home this point.

> '*The failure of self-regulation to live up to its promise is attributable to factors that have, for the most part, been overlooked by its advocates. In their attempts to make over managers' value systems and restructure the modern corporation, they have largely neglected the very real limits on managers' discretion that result from the operation of the market economy. As a consequence of these limits, managers are largely* unable *to consider their firms' impact on society or to subordinate profit-maximization to social objectives, no matter how well-intentioned they are. ... Advocates of social responsibility overlook the extent to which the firm's behavior is a function of market imperatives rather than of managers' values or corporate structure. ... In a market economy, firms are usually* unable *to act in their own collective interests because "responsible conduct" risks placing the firms that practice it at a competitive disadvantage unless other firms follow suit.*' (Maitland, 1985: 132-136)

Maitland's criticism of the notion of self-regulation can most definitely be applied to the argument in favour of co-management, but with the necessary modifications. Some advocates of co-management for example are well aware that their new strategy carries implications for thinking about the market and its internal structure. They seem to be aware that co-management is only feasible on a serious scale if the discipline of the market is somehow mitigated. Jentoft (1989: 144) for instance tries to get round the notion of market discipline by arguing in favour of structures which make agreements between businesses possible at business sector level.

Another modification, but in almost the opposite direction, is to maintain that the discipline of the market in present-day reality is not as strong as Maitland suggests here. Maitland's point would hit its mark most effectively in a market designed according to neoclassical guidelines. However, empirical research in the alternative economic tradition shows that actual markets deviate significantly from this model-based situation (Child, 1972; Dore, 1983; Breiner, 1995: 25-47; Galbraith, 1969; Etzioni, 1988; Granovetter 1992b: Smelser and Swedberg, 1994). Even without extensive research, all kinds of evidence can be found to indicate that market actors definitely do have a certain measure of freedom to act. One clear example is the popularity of strategic management in the business world. The idea of strategy implies choice and therefore scope to act freely. The popularity of corporate social responsibility is another clear indication. This too indicates that market parties have some freedom to act.

Bearing these modifications in mind, we can conclude that it would be short-sighted simply to accuse the advocates of co-management of not taking market discipline into account. On the one hand it is clear that they do pay some attention to market discipline, while on the other hand market discipline in modern-day markets is not so punishing that it does away with all discretionary space in which to act. However, we may reproach the advocates of co-management for not thinking through the relationship between market discipline and citizenship to a sufficiently theoretical extent. A serious exposition on co-management not only has to show how citizenship is possible in the face of market discipline. It must also show that there are ways of mitigating competition which are consistent with the liberal democratic ideal of a competitive order and which do not lead to predictable problems such as the pushing up of prices. Furthermore, it would be convenient if this exposition also indicated how these strategies related to the existing judicial order.

It seems as if Maitland (1985: 145), too, ultimately seeks to put this same spin on his criticism of self-regulation. His argument does not end as a fierce attack on the unifiability of market and collective responsibility. Instead Maitland wonders whether the American market is ripe for consultative structures like those which can often be found in North European nations. In other words, he also assumes that, in one way or another, collective responsibility has to be taken in the market.

Corporatist pitfalls

As stated above, the feasibility of co-management can be criticized on many other points. Instead of delving into these, however, I will go on to examine co-management from a normative angle. Does co-management, regarded as the new offspring of liberal-democratic theory, show enough respect for the tradition into which it is expected to fit? Unfortunately there is much room for doubt in this respect. Co-management appears to pay little heed to key liberal-democratic dogmas. The first of these is the liberal-democratic doctrine that market parties and the state should keep their distance from one another. Advocates of co-management trample on this principle, with their proposition that market parties and the state should get together for consultation on a regular basis. From a liberal-democratic perspective this idea must be rejected. In order to clarify the reasons behind this rejection, I will now give a brief account of the liberal-democratic critique of corporatism. Given the similarities between corporatism and co-management, this criticism can just as easily be applied to co-management.

Corporatism can be defined as the attempt to build bridges between market parties and the state. Corporatists work to achieve a certain measure of cohesion and cooperation between the two sides. The protagonists of corporatism see such cohesion and cooperation (in short: fusion) as both necessary and desirable to ensure that things go smoothly when the state and market parties team up to tackle public issues.

This partial fusion can take on all manner of forms. One example might be an institutionalized consultation in which representatives of market parties speak with civil servants and administrators about plans and future policy. Whatever form is chosen, an essential condition for any corporatist structure is that the negotiators take each other's positions seriously, that they show a certain measure of willingness to compromise and that they have been given some kind of mandate by those whose interests they represent. Structural consultation is a waste of time for the market parties if the state does not use the consultation as a basis for determining policy to some extent. The consultations will be worthless for the state if the market parties are not prepared to take a measure of responsibility for political-administrative issues.

From a liberal democratic perspective, there are a number of dangers lurking within the corporatist vision. A first hazard lies in the fact that, in a society organized according to corporatist principles, the state may gain *direct* influence over market processes. By definition there is no way that this can benefit the rationality of the market. Administrators, by virtue of their position, do not play by the rules of the competitive game. Not only are they less familiar with the working of the market than business people, but they are also insensitive to the disciplinary power of the market. After all, they do not have to answer to the market, but to the political process. Stated concisely, the influence of the state on the rationality of the market will at best be negligible.

It is important to define the significance of this position clearly. When I say that direct state intervention can never work to the benefit of the rationality of the market, I am simply saying that process management is detrimental to the market. This by no means constitutes a plea for an unfettered market. According to the liberal-democratic tradition, the market may be controlled indirectly by limiting its conditions. As I see it, the distance that the indirect responsibility model creates between market and state centres on exactly this issue. And it is on this point that corporatism in practice has the tendency to fall short (Freeman et al., 1973; Peterse, 1990; Yeager, 1991). Corporatism inclines towards process management.

The butter mountains, meat surpluses and milk lakes which dot the landscape of Europe's common agricultural policy exemplify the problems that corporatism may generate in this regard. Another dramatic example from the Dutch context concerns the scandal surrounding the Rijn-Schelde-Verolme shipbuilding concern in the early 1980s. The Dutch Ministry of Economic Affairs provided financial aid and other forms of support to the ailing operation. It cost the state hundreds of millions of dollars but even this was not enough to save the business. Yet the Ministry's civil servants remained blind to this stark economic reality for a long time. Not only did they pour a fortune into what turned out to be a bottomless pit, but they even went so far as to misinform both parliament and the cabinet in their battle to save the shipbuilders.

The second danger of corporatism is the reverse of the first: corporatism always harbours the danger that the influence of the market parties on the state process will be strengthened. Just as the influence of state on the rationality of the market can only end in problems, so too the influence of market parties on the political-administrative process is potentially very dangerous. The state has its own role and responsibilities. These differ from the role and responsibilities of the businessman in the market. In a society where all manner of circumstances conspire to place market parties in a position of privilege, this point cannot be hammered home forcefully enough.

Related to this is the third and perhaps the greatest danger: the threat that corporatism poses to the plural character of liberal-democratic society. Corporatism deliberately seeks to combine the state and market parties, the strongest powers within Western society. From a normative perspective, this is not desirable. The existence of various centres of power is seen in liberal-democratic thought as an important guarantee of the freedom of the individual. And it is precisely because of the lack of opposing forces that a totalitarian society is considered to be the counterpoint to the liberal democracy.

The combining of forces is also undesirable from a public perspective. The liberal-democratic tradition can be said to be a tradition that believes in the value of constructive differences and open non-violent conflict. In the market there is competition, within government there is the battle between various powers and between the political and administrative sections of the state, and in civil society different lifestyles and various political parties jostle for the right to exist. According to liberal-democratic thinking, a well-organized society blossoms as a result of these ongoing conflicts. Modern developments in thinking about rationality underline this belief in the value of difference and open non-violent conflict. Modern authors (Bernstein, 1983; Habermas, 1981) tell us that in a society that has left objectivism and essentialism behind it, rationality can no longer mean anything except 'having sound arguments at one's disposal'. Rationality therefore implies discussion, which in turn implies difference and conflict. A prudent state benefits from an institutional structure that promotes discussion and the battle between sound arguments.

Corporatism, however, does not sit easily with this liberal-democratic ode to open, non-violent conflict. In a corporatist order the negotiators on behalf of the state and those of the market parties are sentenced to one another. This can easily lead to situations in which people are all too willing to paper over differences and embrace compromise. In this situation, mutual respect and understanding transform into an identification with and absorption of each others problems (Selznick, 1949). The risk that the parties within a corporatist set-up might secretly conspire against the public interest must of course never be underestimated.

The administrative development of the Dutch agricultural sector after the Second World War illustrates many of the dangers which accompany corporatism. The friendly embrace which characterized post-war relations between farmers, agricultural companies and public administrators proved to be highly successful in the first few decades. Healthy cooperation between market parties and the state encouraged innovation and the state created all the right financial, infrastructural and other conditions for a far-reaching economic rationalization of the Dutch agricultural sector. Thus began an incredible success story in which the Netherlands, one of the world's smallest countries in terms of surface area, became one of the world's largest exporters of agrarian products.

From the 1970s onwards, however, the disadvantages of such an intimate collaboration became more and more apparent to the outside world. The strategy of rationalization led to concerns about the welfare of the animals, symbolized by the plight of battery hens and calves confined to crates. Enormous environmental problems began to make themselves felt, such as overfishing, pollution caused by insecticides and fertilizers, water depletion in natural habitats and so on. The market parties and the state, who together ran the agricultural sector, refused to recognize these problems for far too long.

In the end it took a collision to make the ship change course. Pressure from society, from parliament and from European agencies forced the Ministry of Agriculture to face up to reality and take a step back from the market parties and the pursuit of rationalization. Now the Ministry has more or less completed its change of position, but the problems in the agricultural sector escalated so much during those years of denial that the solutions are going to require a huge amount of effort. Meanwhile the relationship between the market parties and the state lies in ruins. The farmers, with some justification, feel victimized. After all, it is far more difficult for them to alter their strategy than it is for the state to change tack.

Other political-normative objections

Another liberal-democratic tenet that co-managers also step over too lightly is the notion that, in a liberal democracy, the state has to stay within strict legal boundaries and has to exert its influence by means of limiting conditions. Both these guidelines are very important for liberal democracy. In order to safeguard the freedom of individuals, the state must be kept in check. In a well-functioning liberal democracy, the citizens therefore keep watch on the politicians, the politicians monitor the administrators, the administrators keep an eye on the civil service, while all kinds of other independent organs like the National Audit Office are in place to facilitate such checks. The possibility of running checks is of primary importance in order to keep the state functioning well. After all within this fundamental institution there is no hard hand of discipline as there is in the market.

A further critique on co-management relates to the conflict surrounding the need to make the state powerful while at the same time limiting its capabilities. Here supporters of co-management are too nonchalant in their attitude to putting the state in its place. In this regard I am thinking in particular of the dividing line between private and public domain. This organizational principle is there to ensure that the state does not get a grip on certain domains. This is essential to freedom and the plural character of liberal democracy, yet the advocates of co-management jeopardize this imperative by pursuing cooperation between the state and market parties. The co-management model in effect creates a large grey area between private and public domain. This is something of a paradox. Co-management is ostensibly a way of thinking which argues for a mild form of government, but because it also blurs the boundary between private and public domain, these same mild measures end up indirectly boosting the power of the state.

Another normative reservation I would like to express against co-management has to do with democracy. The advocates of co-management are aware that transferring responsibility from the state to market parties does not exactly go hand in hand with the interests of democracy. Co-management burdens market parties with public responsibility but by the same token it also gives them certain rights, liberties and options within the policy process. In so doing, co-management increases the public power of the market parties. From a democratic point of view this cannot be taken lying down. A way must be found to restore the democratic balance between all citizens. In this respect, the advocates of co-management usually propose pulling up a few more chairs at the table where the businessmen (or their representatives), civil servants and administrators carry out their negotiations. These chairs could be earmarked for certain groups from civil society. I have serious doubts as to whether such a move would be judicious. It poses the threat of creating an even more extensive version of corporatism, with all the fundamental institutions coming together. What is more, simply occupying a chair at the negotiating table offers no guarantee that one will actually be able to make a real contribution to the decisions being made. As I see it, more has to be done on an institutional level to restore the democratic balance.

In conclusion: environmental administrative theory is useful when it comes to providing arguments against the indirect responsibility model. However, even as a practical strategy co-management lacks maturity. It is not well thought out and it does not fit very well into the liberal-democratic tradition.

3.7 LOOKING BACK AND LOOKING AHEAD

For the sake of clarity, allow me to repeat the key objectives of this study. First of all I want to examine the creditworthiness of the indirect responsibility model. Secondly I wish to sketch the contours of an alternative mental model, the direct responsibility model. The operational question I have posed in an attempt to achieve these objectives is: how do various academic disciplines view control over the market in relation to the problems facing the environment. This chapter took a close look at the theory presented by the administrative theorists. What has this examination brought us?

Administrative theory on sustainability, like the administrative thinking of the neoclassical environmental economists, latches onto the debate on the impossibility of managing modern society and the failure of the state in particular. Just like the neoclassical environmental economists, the environmental administrative theorists defend the position that adequate management of the market, if at all possible, will involve changing existing practice and the current mental model of social organization.

However, neoclassical thinking and the thinking of the environmental administrative theorists share nothing else except this general criticism of common practice. For it turns out that the neoclassical environmental economists ultimately seek to *radicalize* the indirect responsibility model. Their ideal is a market in which actors are stimulated to act purely on the basis of market incentives. In contrast, the environmental administrative theorists are out to *replace* this mental model.

Although this study makes use of the critique given by the environmental administrative theorists in order to analyse the indirect responsibility model, this chapter has done more than just present an unquestioning reproduction of their criticisms. A major shortcoming of the administrative theorists' analysis is that its description of the indirect responsibility model is far too sketchy. Another important problem is that the administrative theorists do not make any real attempt to outline the relationship between the model and liberal-democratic thinking. The analysis in this chapter therefore has begun with an attempt to render the indirect responsibility model explicit, to reinforce it and to position it normatively within the liberal-democratic tradition.

On the basis of this attempt we can state that the indirect responsibility model has eight important characteristics.

* *The market is regarded as a sphere in which actors bear a limited responsibility for public issues.*
* *The state has an almost exclusive responsibility for public issues.*
* *The state controls the market by means of limiting conditions.*
* *The state should be a fettered giant.*
* *The distinction between public* issues *and private* issues *comes together with the distinction between the public* domain *and the private* domain.
* *There are not many bridging or linking institutions between the fundamental institutions.*
* *Democracy is interpreted as representation in the state.*
* *Civil society is disregarded when it comes to controlling the market.*

Environmental administrative theorists are highly critical of the indirect responsibility model. They see it as outdated and feel that it no longer reflects the needs of present-day social circumstances. I have divided the main thrust of their criticism into four points:

* *Under present-day circumstances there is a considerable risk that the state will become overloaded in societies organized on the basis of the indirect responsibility model.*

   Overload refers to a situation in which the state has too many tasks on its shoulders, both qualitatively and quantitatively. The result of overload is that public issues remain unresolved. The main circumstance leading to overload is 'the collective dimension of action'.

* *Under present-day circumstances, there is a considerable risk that the state will be confronted with major legitimation problems in societies which are organized on the basis of the indirect responsibility model .*

Due to a multitude of causes, many of which are to do with overload and the collective dimension of action, the state is faced with permanent legitimation problems under present-day circumstances. The legitimacy of the state can be divided into four components: democracy, prudence, effectiveness and *Rechtsstaat*. The legitimation crisis makes itself felt in relation to each of these four components.

* *Under present-day circumstances, the state is unable to use much of its authority effectively in societies organized on the basis of the indirect responsibility model.* The cause of this problem is that, under present-day circumstances, the state is far less powerful than the indirect responsibility model assumes. In many situations, the state is unable to escape from the power of the market parties.

* *Under present-day circumstances, the state does not have the appropriate powers at its disposal in societies organized on the basis of the indirect responsibility model.* The roots of this problem lie in the rival tendencies within the liberal-democratic tradition. Under present-day circumstances, micro planning of the market is essential but at the same time, liberal-democratic thinking is very much opposed to planning. Given its underlying principles, this is a problem that the indirect responsibility model cannot solve.

The above systemization of fundamental problems is based on the arguments as they appear in the administrative literature. For the further analysis of these problems in the following chapters, it is useful to be able to refer to a systematic account which presents a clear analytical distinction between the various problems. First of all we can recognize that, when faced with the problem of declining legitimacy, a distinction has to be made between democracy, prudence and effectiveness on the one hand and the *Rechtsstaat* on the other hand. With regard to the first three points the criticism follows the line that the legitimacy of the state is in decline because its effectiveness, prudence and degree of democracy are diminishing. The criticism in relation to the *Rechtsstaat*, however, centres on the fact that this ideal has its own dark side which can seriously affect the effectiveness of the state. For this reason it makes sense to place the criticism of the *Rechtsstaat* within the framework of rival tendencies. With this in mind, it will not be dealt with as a separate issue from this point on.

A second consideration concerns the loss of legitimacy through diminishing effectiveness. Diminishing effectiveness is partly related to 'everyday' management problems which every large organization has to deal with in modern society. These problems are not relevant within the framework of this normatively inspired study. To a degree the decline in effectiveness is also closely related to inappropriate powers, lack of prudence, democracy and rival tendencies and as such can be better dealt with under these headings. Accordingly this point too will not be dealt with as a separate issue in the rest of this study.

A third consideration under the heading 'legitimacy under pressure' is that it might be better to deal with democracy and prudence separately from one another. After all, they are two separate problems. An increase in prudence does not necessarily lead to an increase in democracy, nor does an increase in democracy necessarily lead to an increase in prudence.

Furthermore, in terms of consequences, the problem of the state being unable to use its powers effectively can be seen as part of its loss of legitimacy due to lack of prudence and democracy. Accordingly this problem too can be dispensed with as a separate issue. Finally it can be posited that the problem of inappropriate powers is closely related to the rival tendencies within liberal democracy. The particular nature of liberal-democratic thinking means that the powers available to the state cannot simply be adapted to the needs of government. For this reason I will characterize this problem from now on as the problem of the rival tendencies. This systemization leaves us with a new list of main problems of the indirect responsibility model:

* *Under present-day circumstances there is a considerable risk of the state becoming overloaded in societies organized on the basis of the indirect responsibility model.*

* *Under present-day circumstances there is a considerable risk that the state will be confronted with major legitimation problems due to lack of prudence in societies organized on the basis of the indirect responsibility model.*

* *Under present-day circumstances there is a considerable risk that the state will be confronted with major legitimation problems due to a lack of democracy in societies organized on the basis of the indirect responsibility model.*

* *Under present-day circumstances there is a considerable risk that no adequate balance can be found between the rival tendencies of liberal-democratic thinking in societies organized on the basis of the indirect responsibility model.*

Despite the fair measure of criticism levelled at the administrative theory, this study ends up underlining this theory's conclusion that the indirect responsibility model is no longer feasible in the present-day context. However, the argumentation behind this position diverges from that of conventional administrative thinking. Here the emphasis is placed firmly on normative and theoretical considerations, which are largely concerned with the doctrine of limited responsibility of market parties and the rival tendencies in liberal-democratic thinking.

Administrative theorists differ markedly from one another in their opinions about the nature of a new mental model of social organization. Some even go so far as to completely deny the desirability of the corrective task of the state as such. The advocates of co-management have no doubt as to the need and the desirability of this task on account of public issues such as environmental problems. This normative choice is an important motivation behind the decision to focus on thinking about co-management as a possible basis for the development of the alternative, direct responsibility model. As it turns out, however, the advocates of co-management fail to develop it as a mental model. Their main focus is on co-management as a practical strategy. Still, a number of prominent characteristics of the direct responsibility model can be inferred from thinking on co-management. These characteristics are:

* *Actors in the market have a certain responsibility for public issues.*

This characteristic forms an immediate and radical distinction between co-management thinking and the indirect responsibility model. The strict division of labour between market and state is thrown overboard in this mental model. Despite their pursuit of profit, actors in the market should concern themselves with certain public issues.

* *The state no longer has exclusive responsibility for public issues.*

    This second point is of course the reverse of the first point. If market parties bear a certain responsibility for public issues, then the state no longer carries this burden alone. In other words, thinking on co-management breaks the equation of state with public domain which is a common characteristic of the indirect responsibility model.

* *The market should be controlled by means of a communicative strategy.*

    According to environmental administrative theorists, the idea of the public responsibility of market parties implies that market parties should consult and cooperate with the state in all phases of the policy process. The state is also required to distance itself from control by limiting conditions. Instead the state should mainly resort to measures which appeal to the voluntary cooperation of market parties.

* *Civil society is rediscovered from a political-administrative point of view.*

    Advocates of co-management involve civil society in their thinking about a new mental model. This is primarily expressed in the notion that parties from civil society should be part of the consultations between market parties and the state.

This short list of main characteristics does not constitute a fully fledged alternative mental model and so we must conclude that the proponents of co-management fail in this regard. But the problems with developing an alternative mental model on the basis of co-management do not stop here. Co-management not only disappoints as a mental model because of its failure to fill in a large number of blanks; in certain respects it actually takes a step in the wrong direction. This tendency manifests itself most clearly in a normative analysis of thinking on co-management, which reveals that it doesn't fit in well with the liberal-democratic tradition. The main objections are that:

* it is not clear how the idea of the market as a competitive sphere can remain intact in combination with ideas about extensive consultation between market parties and between market parties and the state.

* it is unclear how the overload of the state can be dealt with if the state is still involved in the solution to all public issues.

* the idea of communicative management can jokingly be referred to as a liberal-democratic nightmare. This notion does not respect the division between public and private spheres, it is bad for the pluralist character of society and works to the detriment of the internal rationality of each fundamental institution. In other words, it is a one-sided solution to the problem created by the simultaneous and contradictory pursuit of planning and non-planning within liberal-democratic thinking.

* no solution is offered to the problem of limited democracy and limited prudence. Co-management mainly invests in the cooperation between the state and market parties but the likelihood exists that this cooperation will only serve to exacerbate the lack of prudence and democracy problem.

* the reinstatement of civil society has not been systematically thought through.

In conclusion, this chapter has provided a clear picture as regards the first objective. Environmental administrative analyses have provided us with definite support for the claim that the indirect responsibility model is not an adequate mental model of social organization in a society with a large collective dimension. Environmental administrative thinking, however, should be handled with care when it comes to the design of the alternative, direct responsibility model.

In the next chapter, the world of political philosophy is waiting to be explored. There we will investigate the extent to which this discipline is capable of providing the building blocks for a direct responsibility model. At first glance we would appear to have enough reason to set our hopes high: it is primarily the work of the political philosophers which has put civil society on the academic agenda in recent decades. Our look at the administrative theorists suggests that this fundamental institution could be of crucial importance to the design of a new mental model of social organization.

CHAPTER 4

POLITICAL PHILOSOPHY:

Salvation by civil society?

4.1 INTRODUCTION

Within the world of political philosophy, government control of the free market has been a much discussed topic down through the ages. For liberal thinkers at the beginning of the 19th century, such as James Mill, the free market was the Promised Land. Laissez faire was the answer to all questions regarding the management of the economy. But just like Moses, the utilitarian liberals were only able to see the Promised Land far off in the distance. After all, the idea of a free market only achieved its high point in actuality between 1830 and 1870 (Polanyi, 1944: 77 and 83; Searle, 1998; Gray, 1998). The people who experienced this system at close quarters in the course of the 19th century didn't take long to realize that the free market was a long way from being heaven on earth.

> '*Sir Winston Churchill, writing when a young man about the breakdown of the hold of laissez faire on English public opinion, which he dated as occurring in the decade of the 1880's, .. (stated) ... "The great victories had been won. All sorts of lumbering tyrannies had been toppled over. Authority was everywhere broken. Slaves were free. Conscience was free. Trade was free. But hunger and squalor and cold were also free and the people demanded something more than liberty.."*' (Viner, 1960: 68; quotation within the quotation: Churchill, W.S. *Lord Randolph Churchill*. New York, 1906.)

The free market turned out to be a kind of wild animal. It was only fun to get up close if one was protected by a fence. In the unfettered markets where short-term economic interests dictated the course of events, the citizens faced a lamentable fate from an ecological, social and probably also from an economic perspective. In mature liberal democracies such markets are no longer to be found (see Coe and Wilber, 1985: 33). Controlling the market therefore remained a serious political and scientific issue.

This chapter looks at political philosophy. Relatively speaking, political philosophers have given surprisingly little consideration to the issue of controlling the market in recent decades. The indirect responsibility model was fairly common, but it seems that this was an implicit orientation rather than a consequence of deep reflection. One possible explanation for this lack of reflection on social organization is the fact that the 20th century philosophical debate on the adequate control of the free market was obstructed to a considerable extent by the ideological and political struggle between capitalism and communism. This struggle placed the dichotomy of 'plan versus market' at the heart of the discussion (see for example Breitenbach, Burden and Coates, 1991). Given the fact that both total planning and unfettered markets are undesirable, however, this is in fact a false opposition which can only defeat thinking about the social order. The most important questions, such as that of how little planning a society with a free market can get by on, are shoved into the sidelines by this dichotomy (see also Dahrendorf, 1966; Sartoni, 1987 or Geelhoed, 1996:12). The same is true of the discussion on the organization of the welfare state. This was often couched in terms of 'state or market', while the welfare state - as reformed capitalism - actually makes far-reaching demands of both fundamental institutions. Interesting questions therefore always have to do with the relation between market and state (see for example Bowles and Gintis, 1986).

The struggle between capitalism and communism does not tell the whole story, however. Contemporary political philosophers have also tended to show very little interest in administrative and institutional issues (see also Ankersmit, 1995: 207). They have spent decades analysing concepts like 'freedom', 'equality' and 'democracy'. Insofar as questions concerning the social order have captured their attention, this primarily involved evaluation from a narrow normative perspective (see also Cohen, 1995b: 10). Meanwhile, quite a few political philosophers walk into the trap of 'culturalism' which reduces all political-administrative issues to the collective consciousness of the citizens (Van der Wal, 2001; Achterberg and Zweers, 1984; see also Pepperman Taylor, 1997: 88 or Vandevelde, 1992).

But there are exceptions of course. Roughly speaking there are three political-philosophical undercurrents that resist the indirect responsibility model explicitly. The first body of theories is rooted in the radical left and/or radical ecological circles. Advocates of this alternative want to close down the free market to a large extent. Sometimes they argue in favour of a state-run economy (Ryle, 1988). More often, however, they desire an economy which is embedded in and muzzled by local customs, norms and values (Barnet, 1980; Dryzek, 1987; O'Riordan, 1997 or Hoogendijk, 1993). I do not hold out much hope for this alternative, as it means throwing the baby out with the bathwater. Accordingly I will not go into these theories in much depth, limiting my analyses to a short critique in the next section.

The second body of theories is inspired by the direct responsibility model to a greater or lesser extent. These theories certainly do not form a homogeneous tradition, but they all embrace the idea that in the market, actors ought to take some direct responsibility for public issues. Many of them also deem it important to involve civil society in political-administrative processes. Many authors who can be placed in this tradition also strive for a measure of de-differentiation or fusion between the fundamental institutions. In short, corporatist yearnings are most emphatically present within this alternative. Historically speaking, its main identity is as the way, the truth and the life of the Christian thinkers (Pesch, 1928 or Aalberse and Pesch, 1912). But a few liberals and social democrats also made their way surreptitiously across this service road (Durkheim, 1922; Schmoller, 1890). Nowadays, the most flourishing theory that can be placed in this tradition is communitarianism, staunchly defended by the likes of Walzer (1983), Sandel (1982), MacIntyre (1981), C. Taylor (1979), Etzioni (1988) and Selznick (1992).

The third body of theories is the budding tradition that allocates a central role to civil society. Spokesmen for this tradition include J. Keane (1984, 1988a and 1988b), D. Held (1987), J. Habermas (1981, 1989, 1992), J. Cohen and A. Arato (1992), and P. Rosanvallon (1988). Seen from the perspective of the indirect responsibility model, this tradition is the most moderate. The newly emerging theory on civil society only deviates selectively from the indirect responsibility model, the most important deviation being its desire to make civil society relevant in terms of social management.

It is no easy task to provide an in-depth analysis of all these undercurrents in one chapter. It is therefore better to concentrate on one theory or body of theories. The most obvious candidates here are communitarianism and the theory of civil society. It is hard to decide between these two. They probably have more similarities than they do differences, their bad dose of polemic notwithstanding (Nauta, 1999; Kesting, 2001). I have opted for the theory of civil society in this chapter for two main reasons. Firstly, the civil society theorists are closer to arriving at a contemporary interpretation of the idea of civil society (and/or community) than communitarianism is. Communitarians often see no reason to avoid including an appeal to tradition and history in their thinking about community. Civil society theorists seem to have a stronger awareness that the idea of a civil society is in fact a modern concept, which cannot have its roots in the traditional community.

Secondly it is my express desire to remain within the liberal-democratic tradition in this study. The theory of co-management has been found wanting in this respect. It therefore makes sense in this chapter to opt for a theory which has good credentials in this field. The theory of civil society, when seen in this light, is a far better choice than communitarianism. The former remains faithful to liberal democracy on important points, such as its aversion to corporatism. The latter would appear to disappoint in this regard, an impression the proponents of this theory do much to reinforce: many communitarians go out of their way to reject liberal thinking.

Strengthening the direct responsibility model

In this chapter then we will explore the present-day theory on civil society. The structure of this chapter deviates somewhat from that of previous chapters, however. I will concentrate from the very start on the specific question of whether this theory is in a position to strengthen the direct responsibility model. This focus is motivated by the fact that the first central research question of this study, concerning the *necessity* of a new mental model of social organization, has been answered in the affirmative in the previous chapter. The only question that still remains unanswered is the second one regarding the contours of a new mental model. I will therefore investigate in this section the question of whether political-philosophical insights can strip away the weaknesses from co-management, while ensuring that the strengths of this interpretation of the direct responsibility model do not suffer in the process.

Although the theory of civil society is not primarily oriented towards controlling the market, there is no reason why we shouldn't embark upon our investigation with high hopes. The theory of civil society shares a number of articles of faith with the advocates of co-management. To a large extent, civil society theorists embrace the administrative criticism of the feasibility and desirability of an all too powerful state (see for example Keane, 1988c; Cohen and Arato, 1992: 11-15, 23-26, 452-456; or Rosanvallon, 1988: 199-200). Furthermore, in their search for a new mental model that might relieve the state, the theorists of civil society do not surrender to normative relativism. Nor do they trivialize the need to address public issues. Here too the similarities with the programme of the advocates of co-management are notable. There is also another likeness that follows on from this point. Just like the advocates of co-management, the political philosophers feel that civil society has been placed far too much on the outside of administrative processes. Civil society's involvement in the public process needs to be expanded once again. Of course, the civil society theorists have elaborated on this notion in much more detail and have given it a far more central place in their thinking.

> '*We believe there are today important elements of a ... project for retrieving the category of* civil society *from the tradition of political theory. These involve attempts to thematize a program that seeks to represent the values and interests of social autonomy in face of* both *the modern state and the capitalist economy ... . Beyond the antinomies of state and market, public and private, (and)* Gesellschaft *and* Gemeinschaft ... *the idea of the defense* and *the democratization of civil society is the best way to characterize the really new, common strand of contemporary forms of self-organization and self-constitution.*' (Cohen and Arato: 1992: 30)

One last similarity between the civil society theorists and the advocates of co-management is their belief in the power of sound argument. Both camps feel that the greater the influence of rational discussion on political-administrative processes, the greater the chance of sustainable development. Two considerations would seem to play a part for both groups. Firstly, both the supporters of co-management and the political philosophers feel that more deliberation will increase the prudence of the state: the predominance of sound arguments will result in better public decisions. In addition to this, both movements nurture the hope that greater deliberation will also strengthen the will towards ecological modernization. Sustainability is a worthy motive within the liberal-democratic context. As the deliberative character of the administrative processes grows, the authority attached to this worthy motive will also expand.

The belief in 'the supremacy of sound arguments' expresses itself among the advocates of co-management primarily in the emphasis they place on consultation, communication, consensus and compromise between market parties, the organizations representing special interests and the state. The political philosophers express this belief more explicitly. They argue more convincingly in support of the thesis that 'sound argument' can be a powerful force within modern society. To this end they employ both a specific theory of action (the theory of communicative action) and a specific theory of democracy (democracy as the pursuit of a rational consensus). What is more, the political philosophers forge an interesting link between their trust in reason and the need to boost the political-administrative relevance of civil society. As they see it, the rationality of civil society is directed towards conviction and reasonableness. In civil society, action is coordinated by these mechanisms, in the same way that the market is geared towards profitability (Cohen and Arato: 1992: 412 and 429).

This introductory comparison between co-management and thinking on civil society makes it clear that the theory on civil society can indeed be useful when it comes to designing an order in which the responsible market takes on a central role. The political philosophers share a number of important intuitions with the advocates of co-management. Not only that but they make these intuitions explicit and work them out in greater detail. Yet we must not allow ourselves simply to see political-philosophical thinking on civil society and the advocacy of co-management as extensions of one another. By and large, the civil society theorists want nothing to do with the hint of corporatism associated with co-management (see for example Cohen and Arato, 1992: 30 and 40). Meanwhile advocates of co-management are surprised at the tenacity with which the civil society theorists cling to the notion of a market that must be controlled by limiting conditions.

All in all the theory of civil society seems ideal as a stone on which to sharpen the design for a new mental model of social organization. There is agreement but at the same time room enough for creative conflict. In this chapter I will consider whether my hope in this regard is justified. Before doing so, I would first like to focus on environmental (green) political philosophy.

4.2. FAREWELL TO ENVIRONMENTAL PHILOSOPHY

In the previous chapters, the basis for examining the discipline in question was formed by the specific sub-discipline that was geared towards sustainability. This chapter dispenses with that principle, mainly due to the fact that our goal has become more focused. 'Green' political philosophy does not provide much of a helping hand when it comes to strengthening the model of direct responsibility. It has little to offer anyone looking for help on new ways to link state, market and civil society.

In this section I wish to explain my critical view of green political philosophy. Quite a few green political philosophers approach environmental problems from a culturalist perspective. This results in an institutional and political-administrative short-sightedness which renders such theories less relevant to my purposes. What is more, many green political philosophers turn out to be advocates of the indirect responsibility model (for example: Daly and Cobb, 1989; Mulberg, 1995). As such they are equally unable to help us achieve our goal.

Of course, there are green political philosophers to whom none of this applies. To begin with there is a group that Robyn Eckersley (1992: 119-144) refers to as 'ecosocialists' and which counts M. Ryle among its members. At the core of the mental model of the ecosocialists is the notion that the market should be placed under the far-reaching authority of the state (Eckersley, 1992: 122). Sometimes ecosocialists even seem to flirt with the notion of doing away with the market altogether. They feel that the ecological crisis makes an economy led by the state essential. The ecosocialists' rejection of the free market means they have little to offer us. Ecosocialists do not go in search of new links between the fundamental institutions but instead set about getting rid of one of them. What is more, much of the criticism levelled at the indirect responsibility model can be applied many times over to these state-minded thinkers: they seriously overestimate the governmental capacity of the state and underestimate the dangers of a state-run society.

Another group of green political philosophers which is also in search of a new mental model of social organization consists of those authors who mistrust the state as much as they do the market. Sometimes this mistrust springs from the notion that the state together with the market forms an integral part of a capitalist system which bows entirely to the market's short-term rationality. In other cases the emphasis is placed more on characteristics of the state which are regarded as inherently odious, such as its bureaucracy, its hierarchical organization, and its capacity for taking binding decisions (for example Dryzek, 1997: 108-127). According to these authors, simply doing away with the market is therefore not a sufficient condition for achieving sustainable development: the state too must be dismantled. Sustainability will only stand a chance if both fundamental institutions become part of a decentralized society in which citizens united by community lead or supervise social processes in a spirit of healthy consultation. In their eyes, sustainability calls for - what I regard as - a romantic notion of civil society (for example Barnet, 1980; Dryzek, 1987). All will be well if citizens come together to discuss matters and reach agreement.

> '*Once the content of social norms or broad principles for action have been agreed upon, then the creative energies of individuals, in isolation or in combination, could be released to solve problems. Communicative rationalization promotes the cooperation of individuals concerned with different facets of complex problems.*' (Dryzek, 1987: 205)

This mental model I also regard as unsuitable for my purposes. Here not only one but two fundamental institutions are sent packing and once again the advantages of the market are all too easily cast aside. In this case, however, it is even more objectionable that the need for a state is dismissed, especially in a society plagued by a host of collective problems. As far as I am concerned, the great emphasis these authors often place on the decentralization of social processes does not salvage their mental model. Under modern-day circumstances even a small-scale society needs a government to enforce the law and to solve public problems. In addition to this, it needs a state to arm itself against the intended or unintended and foreseen or unforeseen consequences of the actions of other (self-supporting) communities. The fate of tribes like the Yanomano is proof of this. By themselves, the Yanomano stand helpless against the destruction of the South American rainforests that constitutes their habitat.

Participation as solution

One last group of green political philosophers searching for a new mental model resists the temptation to seek a solution in terms of doing away with one or more fundamental institutions. This group champions the idea that democratization of the state and participation of the citizens in political-administrative decision-making is urgently required. To give sustainable development a push in the right direction, a democratic restructuring of the links between state and civil society is essential. The opportunities for citizens to take part in decision-making processes within the state have to be expanded considerably. Voting rights, participation via political parties and other existing means of participation and contribution are not enough. Given that the scale of modern-day societies often gets in the way of effective participation, the argument in favour of democratization and greater participation is often accompanied by a plea for decentralization.

Implicitly, the advocates of democratization appear to differ as to the meaning of participation and democratization (democratization for short). Dryzek (1987) uses democratization to give a new impulse to the ideal of people's sovereignty. The participation of the citizens is the modern-day interpretation of the pursuit of a government by the people. Someone like Paehlke (1989) sees democratization primarily as a way of strengthening democracy as consultation. The citizens do not take control of the government in Paehlke's version of events. This difference notwithstanding, the significance of democratization is the same in all cases: democratization is intended to increase the prudence of the state. Democratization will lead to an increase in and expansion of the deliberative element of political-administrative processes. In addition to this, democratization will create greater openness of government, greater public control and expand the possibilities for ensuring the accountability of the state.

I would like to spend a little more time discussing this theory of democratization. One reason for doing so is because the theory of civil society partly has its roots in a critique on the ideal of democratization. Another is that the idea of democratization and participation is not unique to *green* political philosophers. Participation and democratization can even be referred to, derogatorily of course, as the Pavlov response of political philosophy in general. No matter what public problem is under discussion, political philosophers almost always resort to participation as part of the solution (for example, Barber, 1984; Gould, 1988; Ingram, 1993).

I do not see the theory of democratization, as expressed by the green political philosophers, as a golden gateway in thinking on a new mental model of social organization. For one thing, modern-day political-administrative decision-making processes are complex. They call for a detailed and refined positioning, which demands a great deal of time and expertise from people. This places a limit on democratization beyond which its pursuit will degenerate into populism, or alternatively into a situation where only a small group is actually in power. To take myself as an example, I only have a very sketchy impression of the political-administrative state of affairs with regard to the arts and welfare. I simply do not have enough time to expand my knowledge and insight in these fields.

Secondly, I share the classical objection against the ideal of democratization. Participation by citizens in a modern, complex and large-scale society is all but impossible to organize (see for example Dahl, 1970 or 1989). Even if it were possible, it would lead to laborious political-administrative processes which consume a great deal of time and generate high costs (Mashaw, 1985). The fact that advocates of democratization also make a plea for the decentralization of political-administrative processes does not go far enough to counter this objection. The political-administrative increase in scale is not a facultative phenomenon. It is a result of the increased collective dimension of action, a fact that in turn is closely associated with the development of the market economy and the technologization of society. Reduction of scale therefore tends not to be an option. After all, it is clearly unacceptable to give those living near a nuclear power plant the exclusive right to decide on nuclear power or to leave the conservation issues concerning a natural beauty spot to local residents alone. All citizens are entitled to a say in such matters.

Thirdly, democratization will always remain only a partial solution. The theory of democratization mainly innovates with regard to the link between the state and civil society. The political-administrative problems affecting sustainability, however, are largely to do with the link between state and market.

If I were to conclude my criticism of the theory of democratization here, then I would join the ranks of many critics of participation and direct democracy in acknowledging that the advocates of democratization are right, in a utopian or abstract sense. Such critics admit that participatory democracy is indeed the best form of democracy, but regretfully recognize that it can only work when confined to small city states like ancient Athens (for example Dahl, 1989). I, however, am not convinced of the normative superiority of the theory of democratization. The democratization processes since the Second World War have without a doubt improved the quality of political-administrative processes (Paehlke, 1989). However, this is not to say that a theory which *only* preaches democratization is automatically right from a normative perspective.

For one, a participatory democracy then exposes itself to the danger that the citizens will use their voice or their right of participation to frustrate the political-administrative process. If this happens, then the participatory democracy degenerates into what Mashaw (1985: 29-30) calls *'bargaining at the micro level'*. A government of this kind is no longer prudent but unfit to govern.

Some theories about democratization tackle this problem by placing demands on citizens in a democracy, such as political activism, 'rational' initiative (meaning that people take initiatives while retaining an awareness of the position of others), a disposition towards reciprocity, flexibility, open-mindedness, tolerance, commitment and responsibility (Gould, 1988: 286-299). In my opinion it is very easy to ask too much of the average citizen with such a list of demands (myself included). This renders the theory of democratization utopian in the negative sense of the word.

One could argue against this conclusion with the claim that every modern theory of democracy has to make an appeal to similar characteristics. To a certain degree this claim is valid. Nevertheless it is not enough to salvage the theory of democratization. A typical feature of representative theories of democracy is that they come up with mechanisms, such as control mechanisms and demands with regard to public responsibility, which force politicians and administrators to develop such personality traits. Theories on representative democracy know that virtue has to be squeezed out of individuals.

Another reason to harbour normative doubts about the theory of democratization is the centuries-old but still valid argument that we should not equate government by majority with prudent government. The legitimacy of the state depends both on the degree of democracy it exhibits and its prudence. These two demands are not always equivalent (Cohen, 1986: 27). The possible conflicts between them are, as I see it, not taken seriously enough by the greens in their theory of democratization. A majority can easily decide to persecute minorities or take no account of future generations or animal rights.

A final normative argument against the theory of democratization is that its advocates do not realize sufficiently that our attitude towards democracy has changed in two essential ways since ancient times. First of all, the modern liberal democracy differs from the classical version in the way it values political activity. In ancient times, politics was an important way to spend your life (if you could afford it). In the modern liberal democracy this ideal has lost some of its glory. The appreciation of civil liberties on the other hand has risen considerably. In normative terms, therefore, the theory of democratization places too great a burden on the citizens. The second essential change is that the modern notion of what constitutes a 'political issue' is far wider than the classical. The modern liberal-democratic ideal takes on the responsibility of ensuring a certain standard of living for all citizens. In the classical world, the political domain was far more limited. The economy, for example, was excluded entirely. This normative expansion makes the world of politics considerably more complicated and places boundaries on the possibilities of democratization.

4.3 A PRESENT-DAY INTERPRETATION OF CIVIL SOCIETY

Civil society, along with the market and the state, is the third fundamental institution of the modern liberal-democratic order. So far little has been said about this sphere, as the argument has mainly centred on the indirect responsibility model, a tradition in which the political-administrative role played by civil society is regarded as negligible.

In the sections that follow, I wish to present the emergent theory on civil society. In doing so I will make an analytical distinction between conceptual clarification of the notion of civil society and the political-philosophical analysis of modern society based on the rediscovery of this concept. In this section I will discuss how present-day political-philosophers like Keane, Taylor, Habermas, Cohen and Arato conceptualize civil society. In my approach I will distinguish between the descriptive, the systematic and the functional perspective. In the next section I will discuss how these contemporary political philosophers analyse the condition of modern society in the light of their thinking about civil society.

Civil society from a descriptive perspective

The modern definition of the concept 'civil society' has its origins in the theory on the legitimation of the state, a theory which closely followed the actual rise of the state in modern society. From the end of the 16th century until the middle of the 17th century, the state was defended by prominent secular authors as offering individuals the chance to rise above a natural state of chaos and war. In this context, the concept of civil society was defined as the opposite of the bloody 'natural condition'.

Hobbes's views are typical of this school of thought. In *Leviathan* Hobbes (1651: 84) sketched the *Natural Condition of Man (State of Nature)* as a condition in which

> '... there is no place for industry; because the fruit thereof is uncertain: and consequently no culture of the earth, no navigation, ... no arts; no letters; no society; and which is worst of all, continual fear, and the danger of a violent death; and the life of man, solitary, poor, nasty, brutish, and short.'

The *commonwealth* (1651: 111-115) or *civil estate* (1651: 91) stood opposite this as a peaceful order in which it was possible for life to be pleasant. People could take the step from the natural state to the civil state by drawing up contracts with one another and by calling the person of the sovereign into being. The sovereign was there to ensure order, and indeed was the embodiment of the civil order.

> 'And in him consisteth the essence of the commonwealth; which (to define it,) *is one person, of whose acts a great multitude, by mutual covenants one with another, have made themselves every one the author, to the end he may use the strength and means of them all, as he shall think expedient, for their peace and common defence.*' (1651: 114, passage emphasized in original text).

In Hobbes's thinking, sovereign (i.e. state) and civil society were therefore one and the same. From the middle of the 17th century this thinking began to change slowly. The Hobbesian scheme made way for thinking which put some distance between the state (sovereign) and civil society (commonwealth). Locke (1688) and Scottish Enlightenment figures like Ferguson (1768) typify this school of thought. These authors emphasized that social order (civil society) already existed before the rise of the sovereign. Although this drove a conceptual wedge between state and civil society, authors like Locke and Ferguson did not go on to make an *explicit* distinction between state and civil society (see also Pribram (1983) and Varty (1997)).

During the 18th and 19th centuries thinking on civil society continued on the course set by Locke and Ferguson. The state and civil society came to be more opposed to one another. Thomas Paine was a radical exponent of this distancing process. In Paine's thinking, civil society emerges as a normatively desirable sphere, which ideally should be entirely self-regulating. The state is reduced to a necessary evil that should be kept as small as possible. In Paine's model, modern society is rendered 'uncivilized' by an excess of interference from the state. This is a complete reversal of the Hobbesian position.

Other 19th and 20th century authors also place the state and civil society at odds with one another, although they do not take this opposition as far as Paine did. Most authors would agree that civil society is not a spontaneous order. It needs to be constituted and supported by the state. Nevertheless we can see that civil society as a concept in the 19th and 20th centuries has become more synonymous with 'the non-state sphere'; civil society has come to represent society as a whole, minus the state.

For much of the 20th century, the theory on civil society hardly figured at all in the world of political philosophy. At the end of the 20th century this situation changed dramatically. This was partly due to developments in Eastern Europe, where 50 years of totalitarianism meant that society itself had to be more or less reinvented from scratch. This kindled great interest in the issue of civil society among political philosophers, such as John Keane (1984; 1988a; 1988b).

Keane's work (1988a: 14) with its description of civil society as 'the non-state sphere' carries on the historical development of the concept. In this he occupies an exceptional position. Other present-day political philosophers, like Cohen and Arato (1992: ix and 423-433), Walzer (1991) and Taylor (1991) are out to break with tradition. In their eyes the concept of civil society deserves a new conceptualization in the light of current developments. These writers characterize modern society in terms of three spheres: state, market and civil society. In Keane's description the fundamental difference between these last two spheres evaporates. Everything outside the state is reduced to the market (just as in the work of the neoclassical economists), or conversely: the significance of the market for modern society is lost sight of. This criticism would seem to be justified and should certainly not be ignored in a study about controlling the market.

Cohen and Arato have another reservation about Keane's 'traditional' description. In their opinion the description of civil society as 'everything outside the state and the market' is too broad.

> 'It would be misleading to identify civil society with all of social life outside the administrative state and economic processes in the narrow sense.' ...
>
> 'The differentiation of civil society from both economic and political society seems to suggest that the category should somehow include and refer to all the phenomena of society that are not directly linked to the state and the economy. But this is the case only to the extent that we focus on relations of conscious association, of self-organization and organized communication. Civil society in fact represents only a dimension of the sociological world of norms, roles, practices, relationships, competencies, and forms of dependence or a particular angle of looking at this world from the point of view of conscious association building and associational life.' (Cohen and Arato, 1992: ix-x)

This refinement by Cohen and Arato is also justified, especially as part of an attempt to pin down the nature of civil society within the framework of a political-administrative study. In that context, not all processes and occurrences outside the market and the state are relevant. The cup of sugar I borrow from my neighbour, my weekly drama club rehearsal and the friendship I strike up with another person; these processes and occurrences take place in a sphere that is not of direct political-administrative relevance.

On the basis of these considerations Cohen and Arato arrive at the following description of civil society:

> 'We understand 'civil society' *as a sphere of social interaction between economy and the state, composed above all of the intimate sphere (especially the family), the sphere of associations (especially voluntary associations), social movements, and forms of public communication.'* (Cohen and Arato, 1992: ix)

This description of civil society is along the same lines as the one given by the influential German author Jürgen Habermas:

> 'What we call civil society today no longer includes ... the ... economy. Its institutional core is formed rather by those non-state and non-economic combinations and associations on a voluntary basis which provide an anchor for the ... social components of the lifeworld.' ...
> 'Civil society is composed of those associations, organizations and movements, arising more or less spontaneously, which pick up, condense and amplify the resonance that social problems produce in the private sphere and pass it on to the political public.' (Habermas, 1992: 443)

Cohen and Arato's description of civil society is also the one that I will use in this study. That said, there are a couple of observations that need to be made. The decision to describe civil society as part of the sociological world of norms and rules etc., first of all means that Cohen and Arato need another concept to describe the entirety of social phenomena outside the state and the market. To fill this gap, Cohen and Arato introduce the concept of the 'sociocultural lifeworld' (idem: ix). From an analytical perspective, I find it clearer to distinguish between civil society in the broadest sense and civil society in a narrower sense. The broad version of civil society is everything outside of the market and state. I would describe civil society in the narrower sense as those parts of society outside state and market which are directly or indirectly relevant to the public processes.

In addition to this it is worth pointing out that civil society (in the narrower sense) is not a sphere with territorial boundaries. Civil society is an analytical concept that in reality crosses all kinds of borders. After all, civil society forms a dimension of the social world. It is a cross-section that can only be seen from a certain angle. It is partly for this reason that I think it better to distinguish between broad and narrow civil society rather than the difference between civil society and the sociocultural lifeworld. When I get into a discussion on politics with my neighbour or my friends from the drama club, this can most certainly be seen as part of civil society in its narrower sense. The local knitting club is not part of this narrow civil society if all its members do is knit. But if the chairman invites someone from Amnesty International as a guest speaker, then the knitting club will then fall under the narrower definition of civil society for a time.

Civil society from a systematic perspective

In *Faktizität und Geltung* Habermas (1992: 444) bemoans the lack of clear, systematic definitions of civil society (in the narrow sense). The characteristics of the sphere of voluntary association are unclear. Cohen and Arato share Habermas's view and attempt to fill this gap. In two different ways they try to get a systematic grip on civil society.

In their first attempt, the authors latch on to Habermas's conceptual pair of system versus lifeworld. Although Cohen and Arato are not alone in this, in my opinion it is a strange choice to make. The lifeworld concept is a broad one, that refers more to civil society in the broadest sense of the term than civil society in the narrow sense. In addition Habermas's concept of lifeworld also has an entirely different meaning. For him it also refers to the whole of cultural meaning and background knowledge which enables actors to give meaning to the world.

However, there is a more important observation to be made here: if the critics of *Theorie des kommunikativen Handelns* (1981) are completely unanimous about anything, it is that the conceptual pair 'system and lifeworld' is unworkable from a social-theoretical perspective (see for example: Bader, 1983; De Vries, 1983). These concepts have too many layers of meaning. Each concept refers to a type of action, an orientation for action, a mechanism for coordinating action, a form of social integration, a certain 'reproduction function', and a social sphere. The distinctions that Habermas applies to each of these layers of meaning are in themselves dubious and not very convincing. In addition to this, however, the layers are at odds with one another in many ways. For example, it is not clear why the state - which Habermas sees as part of the system - should not make any contribution to social integration or why, within the rationalized lifeworld, actions should only be communicative. For all these reasons I will abandon Cohen and Arato's attempt to use Habermas's theory to get a systematic grip on civil society. Cohen and Arato also make a different attempt to describe civil society systematically. I will leave the discussion of this second systematic approach until the end of the next section.

Civil society from a functional perspective

Within liberal-democratic thinking, civil society is first and foremost the desirable environment for the good life. To a considerable extent, civil society derives its justification from this contributive normative value. In addition to this, however, civil society is also allocated a number of tasks within the liberal-democratic tradition. To start with, civil society is the breeding ground for modern liberal-democratic society: reproductive processes, personality forming, socialization and culturalization are just some of the processes to take place there. In connection with this study the task that academic researchers refer to as 'social integration' is the most relevant. Accordingly I will devote a little more attention to this theme. What is social integration, what is its importance and why does the task of retaining social integration fall to civil society?

The theory of social integration assumes that a society can only exist in the long term on the basis of shared rules and norms. (see for example: Habermas, 1992). From an analytical point of view, two dimensions can be distinguished within the idea of social integration. On the one hand there is the need for rules and norms for the everyday coordination of activities. On the other hand there is the role that shared norms and values fulfil in enabling individuals to feel that they belong to a community or society .

Leading figures from the social sciences like Emile Durkheim (1922) have invested a great deal in the theory of social integration. Durkheim saw social integration as essential to the survival of a society. Nowadays we see that the importance attached to social integration has been tempered somewhat. Van der Burg (1991) for example wonders whether agreement on ultimate values is always necessary. As he sees it, it is also possible to ensure the existence of a modern, pluralist society if everyone can agree on certain norms and values which each individual justifies for himself in his own way in the light of his own personal absolute values. In addition to this, authors such as Law (1991), Bijker and Law (1992), Harbers (1996) and Harbers and Koenis (1999) point out that solidarity, norms and values are not the only things that bind a society. Societies are also held together by material - by the actual presence of objects. These notions are of course vastly predated by Bentham's (1789) idea that societies are mainly held together by custom and habituation.

Despite all these alternative insights, philosophers and social scientists still agree to a large extent that social integration is a basic condition for society (Polanyi, 1944; Habermas, 1981; Sen, 1987). This theory may even be said to be enjoying something of a revival, given the attention paid to concepts such as 'trust' and 'social capital' in the social sciences nowadays. Even economists, neoclassical and otherwise, have recourse to the subject. Famous in this regard are of course New Institutionalists such as North (1973) and Ouchi (1988). One can also think of the many works which show that trust and other norms and values are a precondition for the effective functioning of the market (Fukuyama, 1995). All this new attention has had at least one positive effect. The theory on social integration is slowly gaining some empirical support. In this regard, Putman's (1997) analysis of regional government in Italy is worth mentioning.

Solidarity, loyalty and shared norms and values are not inherent in man. All of them have to be learned (Nauta, 1999). Most authors who write about social integration see civil society as being of crucial importance to this learning process. According to the defenders of the theory of social integration, civil society is the place where people learn solidarity, norms and values. The other fundamental institutions (market and state) have very little socially integrative power in themselves. Many authors (Schumpeter, 1944; Hirsch, 1977: Habermas, 1981) go even further than this and argue that the market has a negative influence on social integration, breaking down existing socially integrative mechanisms.

This is not the place to investigate the validity of this last statement in depth. It is enough to conclude that there is still a broad consensus for the idea that social integration is important for the continued existence of a society and that civil society plays an important but perhaps not an exclusive role in bringing about this social integration. This justifies the conclusion that civil society is therefore of great importance to modern differentiated society. On the one hand people receive their normative education there. They learn to conceive of themselves in a normative sense as part of a society. On the other hand they learn the specific dispositions, normative and otherwise, needed to be able to function in the market, the state, and of course in civil society itself.

The close relationship between civil society and norms and values often gives rise to the thought that civil society must be a traditional remnant of society. As I see it, this notion is mistaken. I will counter it by presenting Habermas's criticism of Weber, which centres on this point to some extent. I will preface this with a brief excursion into Weber's theory of the modernization of Western society.

Weber sees the history of the Western world since the Middle Ages as a process of modernization. At the heart of modernization lies a process of rationalization which has taken place on both a cultural and an institutional plane. The rationalization of culture mainly entails the creation of a culture that is strongly geared towards controls, towards a systematic and planned approach to actions, towards efficiency and towards instrumentalization of social phenomena. As a result, Western society has become secularized and demythologized. Science and scientific methods have taken the place of belief systems and metaphysical worlds. The most important implication of the rationalization process on the institutional plane is social differentiation, the process in which market, state and civil society not only develop into relatively independent spheres but also become more internally specialized, that is to say geared more specifically towards their own logic and tasks.

Weber looked upon the rationalization process with mixed feelings. In keeping with the spirit of his age he felt that thinking about factual matters could be rationalized but that rationalization of thinking on norms and values was not possible. As he saw it, belief in the rightness of a value or a normative principle always rested by definition in irrational tradition or religion. This makes the rationalization process necessarily a threat to the social integration of modern Western society. While rationalization allows the market and the state to focus ever more closely on their respective ends, civil society ends up losing its logic. Looked at through Weber's eyes, rationalization eats away at the vitality of cultural processes by definition. The cultural processes and cultural learning processes borne by civil society become disrupted. Rationalization teaches people to become sceptical about religion and superstition. Meanwhile it cannot deliver an alternative cultural and normative framework, since all norms and values are irrational.

Habermas is one of the modern-day authors who attempts to crack open Weber's sombre view of modernization. Since Weber's pessimism about modern society is so closely related to his epistemology, it is logical that Habermas sets about dismantling this component of Weber's thinking. Habermas (1981: 25-101) contends that it is possible to reach agreement on norms and values without having religion or tradition as a basis. It is most certainly possible to rationalize the normative discussion.

I will give a brief sketch of Habermas's argument for this thesis, whereby I will also refer to one of his most important interpreters, Richard Bernstein (1983). Bernstein's theory, in my opinion, is more persuasive than that of Habermas himself. Bernstein is completely free of Habermas's tendency to support his theory by recourse to the transcendental. By this I mean that Habermas, in utter contradiction of the spirit of his own theory, sometimes tries to deliver his arguments too forcefully. He attempts to persuade his audience by underlining that his way of thinking is 'inevitable' and that its validity can be understood *a priori* by analysing the structure of language itself (see also Bernstein, 1983: 182-193). Bernstein's approach is more modest and pragmatic. He tries to convince reasonable opponents on the basis of an analysis of the way in which we understand the concept of rationality.

However, let us begin with Habermas. According to Habermas, rational action first of all consists of alter's ability to understand the type of claim that ego wants to express in a given statement. Secondly, rationality implies that alter is able to determine the validity of a statement. Statements can be divided into two main types: statements about facts and statements about norms and values. In the case of statements about facts, determining the validity of a statement implies determining truth or falsehood. In the case of normative statements it involves determining rightness or the lack of it.

It is safe to say that Weber would have been in agreement with Habermas (and Bernstein) up to this point. After all the distinction between norms and facts is crucial to Weber's epistemology. Where the path of Weber (and the positivist tradition along with him) diverges from that of Habermas and Bernstein, is with regard to the question of which statements can be rationally assessed within their 'sphere of validity'. According to Weber, in the sphere of truth there exist rational methods for determining the truth of a statement. More particularly, science has developed methods for the objective and absolute assessment of factual statements. Rational methods of this type, argues Weber, are by definition not available in the sphere of normative validity. There statements remain subjective and relative.

Partly inspired by Habermas, Bernstein criticizes this Weberian position, using both negative and positive arguments. His most important negative argument is that present-day scientific-philosophical research by the likes of Feyerabend (1978) has shown that there are no objectively and absolutely valid scientific methods. Each method is based on normatively charged background knowledge. In principle, therefore, there is no distinction between statements about facts and normative statements in this respect.

Bernstein's positive argument is that we can escape the looming conclusion that *all* types of statements are irrational by reinterpreting the concept of rationality. According to Bernstein the development of modern society since Descartes has been coupled with a specific attitude towards rationality. A rational statement within this specific attitude came to be equated with an objective and absolutely valid statement. The demand that rational knowledge had to be *objective* meant that man as a knowing subject did not play a role in the creation of the knowledge. Opposed to objective knowledge we have subjective knowledge. This is knowledge with regard to which man as a knowing subject is not transparent. Man mediates in the creation of this knowledge in one way or another.

The position that rational knowledge is *absolutely* valid implies that such knowledge is valid under all circumstances. Opposed to this we have relative knowledge, that is to say knowledge that is dependent on context.

According to Bernstein the demands attached by modernity to the possibility of valid statements were far too strict. In this regard he speaks of a Cartesian fear of context-dependent knowledge and knowledge mediated by subjects. What is more he sees these demands as not reflecting the nature of human existence. The only way in which people can determine the validity of statements is by weighing up sound arguments in relation to one another through consultation with others. In Bernstein's vision, rationality is invested with an intersubjective and procedural character (in this too he follows Habermas). Whoever seeks to act rationally must not embark on a lone quest in search of objective and absolute methods. Rational action means creating a context in which only sound arguments count (as opposed to force etc.) and making sure that in this situation statements are opened up to the criticism of others. Stated simply, rationality means only allowing oneself to be led by the power of sound arguments.

By virtue of this argumentation, Bernstein can protect Weber against Feyerabend's cynicism regarding the issue of whether rationality exists at all. At the same time, however, he is also able to protect normative reason against Weber's thinking. Rationality as 'the presentation of sound arguments' can just as easily apply to the sphere of normative validity as to the sphere of validity in which facts are assessed.

It is important to note here that Bernstein is certainly not out to claim that all knowledge is subjective and relative. He considers thinking in terms of the objective-subjective and absolute-relative dichotomies as a legacy of Cartesian fear. To his way of thinking, valid statements are neither subjective nor objective; neither absolute nor relative. They are intersubjective. Man is incapable of knowing reality as it is. All valid statements are mediated by the possibilities that we have for arriving at knowledge. This precludes being able to formulate objectively valid statements. According to Bernstein this does not, however, mean that any claim automatically acquires the status of a valid statement. If ego claims that dogs can fly, this argument only becomes valid if alter agrees with the arguments that ego has presented to support this claim, having confronted ego's statement with his own knowledge and his analytical skills and the knowledge and analytical skills that society has stored in its traditions. A comparable argumentation applies to the distinction between absolute and relative. Rational action is essentially intersubjective.

The upshot of this inescapable digression into the epistemological theories of Weber, Habermas and Bernstein is that the rationalization of society does not necessarily constitute a threat to social integration. A rationalized society can reach agreement on norms and values. Civil society can therefore also fulfil its functions in a modern society. According to Habermas, however, this is no easy task. The normative discussion can only be rationalized under specific conditions. The most important condition in this respect is that processes within civil society must be coordinated in a specific way. A rational normative discussion calls for a civil society in which actions are coordinated on the basis of sound arguments, discussion and open processes of consensus-forming. Habermas calls this kind of civil society a 'modern civil society'. In such a civil society there is no room for money, power and prestige based on tradition. The modern civil society is the sphere of Habermas's famous 'domination-free discourse', in which only 'communicative rationality' counts. From this perspective it is easy to understand why Habermas calls voluntary associations the institutional core of civil society. In Habermas's view, voluntary associations provide suitable institutional bedding for domination-free discourse, since he sees them as mainly democratic organizations, where no one is forced to participate against his will.

Given Habermas's reinterpretation of rationality, it should come as no surprise that his appreciation of civil society turns Weber's vision on its head. Weber looked upon civil society as the sphere of tradition and religion-based norms and values and therefore the most irrational sphere. Habermas and Bernstein regard civil society potentially as the most rational sphere. In its modern form, civil society is the only sphere where sound arguments determine the coordination of action. Rational argument is not obstructed by the pursuit of profit (as in the market) or power and hierarchy (as in the state).

A number of authors, including Charles Taylor (1991: 52) express reservations about Habermas's assertion that modern civil society is the most rational sphere. Such reservations are relevant to the present argument because several other key authors on civil society, Cohen and Arato among them, follow Habermas in this line of thought. Taylor does not see how Western civil societies as they actually exist give a basis for claiming that they are completely rational spheres in which processes are regulated on the basis of sound arguments. Existing civil societies are often under the influence of prejudices, governed by practices determined by tradition and so on. As Taylor sees it, Habermas's position on the rationality of modern civil society refers to an ideal that has not been realized.

In response to this critique of Habermas, and indeed of themselves, Cohen and Arato (1992: 456) claim that Habermas's theory is both empirical and normative at the same time. They refer to it as a 'two dimensional concept', arguing that the concept of civil society carries both an empirical and a normative connotation. On the one hand it points to civil societies as they actually exist. On the other hand it is a normative concept which stipulates what a modern, rationalized civil society should look like. Arato and Cohen not only see this duality as illuminating for Habermas's theory. They regard it as part and parcel of *all* knowledge and insight that is inspired by the liberal-democratic tradition: the pursuit of a better social order being its essential characteristic.

In my opinion, the idea of the two dimensional concept is simply not enough to save Habermas, and Cohen and Arato. This way out is too easy. When wielding a two dimensional concept, one has to know exactly when each of the two meanings applies. Habermas, Cohen and Arato are not especially conscientious in this respect. This results in a skewed, overly normative picture of civil society. One example is the notion shared by all three that voluntary organizations within civil society are democratically organized, and that their activities are coordinated on the basis of communicative action. From an empirical point of view, nothing could be further from the truth. Non-governmental organizations, from Greenpeace and Friends of the Earth to Amnesty International are not renowned for their concern with internal democracy.

4.4 CIVIL SOCIETY AND PRESENT-DAY SOCIETY

In this section I will discuss how the flourishing theory of civil society has placed this concept at the centre of a critical analysis of modern-day society. Given the specific aim of this chapter, I wish to concentrate on analysing a single theory in great detail. The analysis should be as specific as possible, after all, in order to highlight the contribution that civil society theory can make towards reinforcing the direct model of responsibility.

Looking around for a suitable candidate, the work of Jürgen Habermas presents obvious possibilities. Habermas did a great deal of work as part of the philosophical project to 'save' practical (or normative) reason. However, I would like to leave Habermas be, since I judge his theory to be rather weak when viewed from a political-administrative perspective. In support of this claim I will give a brief criticism of *Theorie des kommunikativen Handelns* (1981) and *Faktizität und Geltung* (1992).

One of the main reasons why *Theorie des kommunikativen Handelns* is not particularly interesting from a political-administrative perspective has already been given: 'system and lifeworld', the concepts which Habermas uses in this study to conceptualize social reality, lack validity. A second, closely related reason is the one-sidedness of Habermas's analysis of the state. As Habermas sees it, the state is part of the system, nothing more. It serves only the market and itself, and it is focused exclusively on power. Approaching reality via this conceptualization simply does not work, however, since the state clearly has not severed all ties with the citizens and civil society. To some extent at least, it must be seen as a democratic organ, one that regulates collective issues for the citizens. In my view, this is how citizens in liberal democracies continue to experience the state, at least to a certain extent. The turnout figures at elections provide ample proof of this. We can also say that Habermas is wrong to limit his analysis of the state to the functional perspective. The state must also be examined from a hermeneutic perspective, since it cannot function if cut completely adrift from the lifeworld of the people, at least not for very long (see also Walzer, 1991: 301; Durkheim: 1922).

The third reason for rejecting *Theorie des kommunikatieven Handelns* follows on from this. Habermas's conceptualization leaves him next to no room to solve the problems with regard to controlling the market. The market and the state are inaccessible for the citizens, if they are governed in a purely systemic way. The citizens of Habermas's rationalized lifeworld are trapped in a classical tragedy. They are able to see that the 'project of the Enlightenment' (freedom, equality, emancipation) has run aground, but they are not able to do anything about it (Cohen and Arato, 1992: 389-410).

In order to salvage some of the remains of the Enlightenment's ideals, Habermas proposes setting civil society further apart from the market and the state. In other words, system and lifeworld have to be separated more emphatically. This pursuit of an absolute separation to me constitutes a disastrous headlong escape. However autonomous or independent civil society may be, in the modern world it does not exist on an island. Processes within the market and the state are relevant to this sphere, if only because they have consequences, intended or otherwise, which exert an influence. Cohen and Arato add that this pursuit of absolute separation seems unnecessarily radical, especially when it comes to the state.

> '*The need for* (specific - wd) *steering mechanisms for the state and for the economy must be respected if we expect them to function efficiently. This, as is well known, mitigates against total democratization along the lines of direct participatory models. Yet it would be fallacious to conclude that no democratization is possible in these domains.*' (Cohen and Arato, 1992: 415)

Faktizität und Geltung (1992) can be interpreted as Habermas's political-administrative revenge. It is one of Habermas's later works and in it he presents a more effective way of distinguishing between the three fundamental institutions. He also places the state less exclusively within the system. In *Faktizität und Geltung* the state partly serves civil society, contributes towards social integration and is also, among other things, a vehicle for achieving liberal-democratic objectives.

Yet still Habermas fails to obtain a proper feel for the state and its position. In *Faktizität und Geltung* two conceptions of the state vie for supremacy. One is the interventionist or tutor state, which Habermas treats with the same deep mistrust that he reserved for 'the system' in *Theorie des kommunikativen Handelns*. Although Habermas sometimes recognizes that the need for intervention, micro planning and process management might sometimes arise from a desire to materialize the ideals of a liberal democracy, he feels that by definition it can only have the opposite effect. Intervention and state tutelage undermine solidarity, social integration, autonomy and, more generally speaking, civil society.

The other conception of the state that features in *Faktizität und Geltung*, is the state which restricts its actions purely to control by limiting conditions. Habermas does not mistrust such a state. In fact, he even seeks to strengthen it. Unfortunately a theory with such a normatively charged dual vision of state can be of no use to us in this study. As I have already described, the conflict between the pursuit of government by limiting conditions and the actual need for micro planning is a paradoxical tendency within the liberal-democratic tradition. Habermas settles the battle between these rivals one-sidedly. Given the collective dimension of action this approach does not help matters.

The theory of Cohen and Arato

It is partly as a result of this political-administrative-oriented criticism of Habermas that Cohen and Arato enter the picture. Cohen and Arato teamed up to write *Civil Society and Political Theory* (1992). One of their motives for doing so was to perform a kind of political-administrative makeover on Habermas's theory. I am certainly not alone in thinking that the authors succeeded in their aim. Nowadays, *Civil Society and Political Theory* is widely regarded as one of the most comprehensive theories of civil society (see for example Habermas, 1992: 444-445). All the more reason then to devote considerable attention to the theory developed by Cohen and Arato in this section.

In order to gain access to Cohen and Arato's line of reasoning I would first like to delve into their background and their motivation. In normative terms, Cohen and Arato position themselves as firm supporters of the 'radical-democratic tradition', a tradition which also includes authors like Gould (1988), MacPherson (1973 and 1977) and Pateman (1985). Jean Jacques Rousseau is often referred to as the patriarch of this tradition, which distinguishes itself from other movements within liberal democracy by virtue of the great emphasis it places on autonomy as an ideal. Within this framework, its supporters argue in favour of each citizen's right to a considerable degree of positive freedom.

There is something paradoxical about the pursuit of autonomy within the context of a *society*. Living together as part of a society is only possible if people are prepared to subject themselves to rules. Each rule however, constitutes a potential threat to the freedom of the individual. The liberal-democratic tradition usually gets round this paradox by stating that an individual is autonomous if he only has to subject himself to rules he has chosen himself.

A typical feature of the radical-democratic variant of liberal-democratic thinking is its rather literal interpretation of the idea of subjecting oneself to self-chosen rules. In a 'real' liberal democracy citizens actually participate in government on a large scale. In the eyes of the radical-democratic tradition, a true democracy is a participatory democracy. Cohen and Arato no longer believe in this ideal, which lies so close to the utopia of a direct democracy. According to them, direct democracy is an illusion and most people nowadays are aware of that (see also Section 4.2). Cohen and Arato therefore feel that the radical-democratic tradition can only maintain its appeal if it is modernized. To begin with this means that the tradition has to dispense with what I wish to call 'the polis syndrome' of radical democracy. According to Cohen and Arato, in complex modern-day society, the idea that the classical Greek polis is the template for democracy has to be buried for good. In this connection, Arato and Cohen also feel that the radical-democratic tradition should finally renounce its secret longing for a dedifferentiated society. As they see it, every sensible theory of democracy takes the differentiation of society in state, market and civil society as a starting point.

Despite all this criticism, however, Cohen and Arato are emphatic in their insistence that a departure from direct democracy does not necessarily imply that one has to reject the ideals of the radical democracy. Instead, Cohen and Arato seek to modernize the radical-democratic tradition.

> '*Our approach differs from theirs* (the advocates of direct democracy - wd) *in arguing for more, not less structural differentiation. We take seriously the normative principles defended by the radical democrats, but we locate the genesis of democratic legitimacy and the changes for direct participation not in some idealized, dedifferentiated policy but within a highly differentiated model of civil society itself.*' (Cohen and Arato, 1992: 19)

In other words, there is a need for a new, modernized radical-democratic theory of democracy and Cohen and Arato have set themselves the goal of developing one. Normatively speaking, this alternative should be able to rank alongside the classical radical theories of democracy. At the same time it should be better constructed from a historical, systematic, institutional and political-administrative point of view. In this regard they call particular attention to the issue of linkage between state, market and civil society. According to Cohen and Arato (1992: 19) this issue of linkage is in need of urgent attention within the theory.

A second goal Cohen and Arato hope to achieve in *Civil Society and Political Theory* is to demonstrate that the concept of civil society should take centre stage in a modern radical-democratic theory of democracy. In modern society, the ideal of autonomy can only be realized in civil society. A last aim of the authors is to build a bridge between the radical-democratic tradition of democracy and the more elitist, pluralist tradition. As Cohen and Arato see it, the long-standing conflict between these traditions is, in the end, utterly unjustifiable. Both traditions contain elements which are weak and some which are valuable. Cohen and Arato (1992: 2 and 18) see their theory as a synthesis that brings together the strengths of both traditions.

The diagnosis: Habermas's colonialization thesis

Cohen and Arato (1992: 18) regard environmental problems as one of the major social problems of our time. Despite this conviction, they do not embark upon a specific analysis of environmental problems. This has to do with the intellectual tradition in which the authors find themselves. This tradition is often described as the Frankfurt School of Thought (and which is also home to Horkheimer, Adorno and Habermas) is set on coming up with an all-embracing social-theoretical diagnosis at macro level. Concrete social problems like environmental pollution, distributional issues or alienation are seen in the context of this global diagnosis as 'pathologies' which can all be traced back to the same cause or social constellation. Therefore, any inquiry into the nature of Cohen and Arato's diagnosis of environmental problems inevitably leads us to their general portrait of our age. Does modern-day society suffer from pathologies and if so, how can these be explained? Cohen and Arato do indeed feel that modern-day society is far from being free from such ailments. Environmental problems, poverty and the widespread incidence of psychological illnesses like stress are but three examples. In their explanation of these pathologies the voice of Habermas can be heard clearly and unmistakably. Cohen and Arato's diagnosis can even be said to be surprisingly faithful to Habermas. All of which prompts me to give a brief sketch of Habermas's diagnosis of modern society.

As already indicated above, Habermas sees it as his task to correct the pessimism expressed by Weber. Weber stated that in modern society the possibility for rational normative discourse is on the wane. On closer examination, this position can be divided into two separate claims. The first is historical-empirical in nature and states that 'practical reason' is actually under increasing pressure in modern society. The second is analytical-epistemological in nature and states that this historical process occurs out of necessity, because rational discussion of norms and values is impossible.

Habermas argues that practical reason and civil society can be rationalized, therefore he opposes Weber's second claim. He does not object to Weber's first claim, however. Just like Weber, Habermas is of the opinion that practical reason and the vitality of civil society are under pressure in the real world. The historical process of modernization has indeed taken a paradoxical turn: in effect the Enlightenment is in the process of destroying itself (Habermas, 1981, Volume II: 277 and 470). According to Habermas, however, this historical fact is not an inevitable consequence of the rationalization of Western society. The pathologies could have been avoided. The rationalization process could have taken a different turn and, perhaps more importantly, it is still possible to steer it in a different direction.

Habermas then argues that we should not see the 'diseases' from which modern society is suffering as a consequence of rationalization as such. They are instead the result of the unfortunate turn taken by the rationalization process. What has gone wrong, according to Habermas, is that the rationalization process has not occurred at the same pace in each of the three spheres. While rationalization in the market and the state progressed more or less unhindered, the rationalization process in civil society faltered and stalled. Habermas argues that these problems cannot be seen separately from the stormy developments within the market and the state. Both these fundamental institutions exhibit what might be referred to as imperialistic tendencies. In other words, when processes dominated by the rationality of the market or the rationality of the state interfere with processes which are regulated according to another rationality, both have a tendency to force that other rationality into the sidelines. Funeral customs in the Netherlands provide an example of this. Well into the 20th century, a funeral was always organized by the deceased's neighbours. This so-called 'neighbourly obligation' was therefore of great importance. Nowadays funerals are the province of the market.

The dynamics of the market and the state are therefore no respecters of institutional boundaries (Cohen and Arato, 1992: 440 or Habermas, 1981, Part II: 278). Civil society is the main victim of this, as communicative rationality is pushed aside by power and money. As Habermas sees it, this is not necessarily a bad thing in all cases. It only becomes so if money and power take over the coordination of processes within civil society which are essential to social integration. In such instances communicative rationality cannot be pushed aside without inflicting damage. This is exactly what has happened, however, with the result that the rationalization process, which held such abundant promise with regard to freedom and autonomy, has come to have the opposite effect. Habermas characterizes this process as the 'colonialization of the lifeworld'.

> 'Systemic mechanisms ultimately also displace forms of social integration in those fields where symbolic reproduction of the lifeworld is at stake. The mediatization of the lifeworld then takes the form of colonization.' (Habermas, 1981, Part II: 293)

In Habermas's view, colonialization of civil society by the state expresses itself in the form of juridification, bureaucratization and monetarization. The colonialization of civil society by the market primarily takes the form of monetarization. Generally speaking, the negative consequences of colonialization are loss of freedom, loss of autonomy and loss of meaning. An example of colonialization with direct political-administrative relevance is the use that the state sometimes makes of subsidies to obtain influence over non governmental organizations. In 1997, for instance, a number of Dutch MPs threatened that they would try to stop the subsidies given to the Dutch sister organization of Friends of the Earth (*Vereniging Milieudefensie*) when the organization staged protests against the construction of a new runway at Schiphol, by far the Netherlands' most important airport.

It is remarkable how faithfully Cohen and Arato adopt Habermas's colonialization thesis, especially given the reception accorded to the *Theorie des kommunikativen Handelns*. Habermas's thoughts on colonialization are not generally seen as 'the blue monster's' finest hour. The main doubts expressed concern the scope of the colonialization thesis (see for example De Vries, 1983). The idea that *all* social problems can be conceptualized as colonialization of the lifeworld is not widely shared. Environmental problems are often cited as an example in this respect, since they primarily affect the natural world and not civil society.

In addition, Habermas's conceptualization forces him to use the 'colonialization' label to refer to all activities emanating from the state in the direction of civil society. This position is untenable. Modern civil society is not a natural sphere. An adequate theory therefore has to be able to indicate how activities (by the state) which strengthen civil society can be distinguished from activities which undermine it. It must be possible to distinguish theoretically between the granting of formal rights to citizens by the state and attempts by the state to impose its will on these individuals. Nor is it the case, for example, that every attempt by the state to influence the social debate is suspect. The state has a constructive role to fulfil in the process of opinion-forming and consensus-forming. After all, the state has a great deal of knowledge at its disposal where political-administrative issues are concerned, and the public debate can be enriched if such knowledge is made relevant.

There is not enough space here to embark on a far-reaching analysis of the criticism and possible refutation of Habermas's thoughts about colonialization. What is more, there is no need for such an analysis, since in this chapter I am mainly interested in the political-administrative thinking of Cohen and Arato. As will become apparent, these authors attach great political-administrative importance to civil society's status as an independent sphere, dominated by communicative action. With this in mind, I will therefore interpret 'colonialization' primarily as the erosion of this characteristic of civil society. As I see it, this narrower definition of colonialization makes the idea less objectionable.

The therapy: beyond interweaving and detachment

Cohen and Arato may well rely heavily on Habermas for their diagnosis, but when it comes to proposing a cure for society's ills, they do nothing of the sort. Habermas ultimately becomes so ensnared in his own concepts that he is no longer able to develop an effective remedy for the rationalization process that has spun out of control. As he sees it, none of the fundamental institutions are capable of keeping the rationalization process on track.

According to Habermas, the market lacks all the mechanisms necessary to perform such a collective and public operation. It can only be steered by external pressure. With regard to the state, Habermas accedes that in principle the necessary mechanisms for this purpose are present there. Nevertheless, he feels that we cannot expect much of the modern-day state. The state has become a reactive organization which is embroiled in the 'system'. Translated into the terminology of the previous chapter, Habermas is saying that, in the modern-day state, the balance between the idealistic and strategic component of government has tipped definitively in favour of the latter. The state is now exclusively strategic in the way it operates. Civil society also has little consolation to offer in Habermas's thinking. The potential for rational action within this fundamental institution is indeed large enough, but the 'system' (market and state) has become impenetrable for the 'lifeworld' (civil society). Habermas's resulting rescue mission involves making the best of a bad situation. He proposes detaching civil society from the market and the state as much as possible. As I have already suggested, this does not seem to be a very fruitful path to take in a society where interdependence is the name of the game.

Cohen and Arato feel that Habermas's pessimism springs from a stalemate that is essentially the product of his own thinking. They attempt to break this deadlock. Ways must be found to turn the tide of colonialization. Translated into my vocabulary, we could say that the idealistic component of government needs to be strengthened. Within the state, a better balance must be found between the strategic and the idealistic component of government.

According to Cohen and Arato, the first thing needed to achieve this purpose is the creation of opportunities for undisturbed rational discussions, which would enable citizens to find out which public objectives and policies are rational and desirable in the light of liberal-democratic ideals. Secondly, strengthening the idealistic component of government implies that the outcome of these rational discussions should somehow influence the actions of the state. If this influence is great enough, the balance between the idealistic and the strategic aspect of government can be restored.

Where might such discussions take place and how can they exert influence on the state? Cohen and Arato, following in Habermas's footsteps, feel that a new domain has developed within modern Western society, partly as a result of the modernization process. They refer to this domain as 'the public realm'. In the present context, this sphere can be neatly described as the sum of the places where rational, public discussion can take place undisturbed. According to Cohen and Arato, this public realm exists almost entirely within civil society. They see this as no coincidence, since modern civil society is the only sphere where action is coordinated on the basis of communicative rationality. In addition to this, civil society is the only sphere where actors are sufficiently insulated from daily necessities, administrative pressure, strategic considerations and their interests as market actors to take part in such a discussion openly and without disturbance.

Along this line of reasoning, Cohen and Arato do indeed make civil society highly relevant for the government of modern society. Civil society is the best place to rationally discuss public issues and public policy. Nevertheless, 'undisturbed discussion within civil society' alone is not enough to strengthen the idealistic component of government. Civil society also has to exert an influence on the state, not only with a view to passing on the upshot of these rational discussions, but also in order to counter the forces that nail the state to the strategic component of government. Cohen and Arato identify various strategies for increasing civil society's influence on the state. Their trust in the power of communicative rationality is expressed in their belief that public discussion can in itself have an influence on actors within the state.

However, Cohen and Arato do not want to pin too much hope on the power of communicative rationality. Arguments need to be reinforced by institutional power. When turning their attention to this question of power, the authors do not turn their backs on their radical-democratic roots. The main strategy they propose is the democratization of the state, which they see as going further than just the right to vote and the various avenues of consultation already open to the citizens in civil society. However, in my opinion, the issue of exactly what this democratization should involve and what form it should take is not the strongest aspect of *Political Theory and Civil Society*. It is clear that Cohen and Arato see existing possibilities (essentially the ballot and various forms of consultation) as insufficient. It is also abundantly clear that they want greater participation by citizens in administrative processes. However, the authors remain vague when it comes to practical measures for making this happen.

The vagueness that descends on Cohen and Arato's theory at this point may have to do with the fact that democratization serves several ends in their thinking. The need for civil society to influence the state is only one of these. Democratization according to Cohen and Arato is also important from a normative point of view: if autonomy is the goal of the liberal democracy, then Cohen and Arato posit that people should be given a certain say in processes which influence their lives. Cohen and Arato also attach importance to democratization for long-term strategic reasons. In the long term, communicative rationality (and with it the democratic nature of civil society) cannot remain at the required level if civil society is the only fundamental institution where democracy, communicative rationality and rational consensus-forming exist.

> 'Marx made the point long ago, that if democracy is restricted to one sphere ... ((in Marx' case:) ... the state) while despotic forms of rule prevail in the ... (other spheres in Marx' case: the) ... economy and ... (the) ... civic associations), then the democratic forms of the first sphere become undermined.' (Cohen and Arato, 1992: 415)

The emphasis that Cohen and Arato ultimately place on participation and democratization may give rise to a degree of scepticism given the aim of their project. They could be accused of expelling the classical radical-democratic ideal through the front door, only to smuggle it back in again through the back door. I do not agree with this assessment, however. The authors emphasize that the scope for democratizing the market and state is limited. Sooner rather than later, the pursuit of democratization will clash with each fundamental institution's own drive to organize its sphere in accordance with its own logic. When that happens, say the authors, the pursuit of participation will be restricted to treading water (Cohen and Arato, 1992: 415, 468 and 479).

Self-limiting utopia

Analysing Cohen and Arato's theory has so far given us the opportunity to distinguish between three kinds of theories. To begin with there is the 'classical' radical-democratic theory which Cohen and Arato are seeking to rejuvenate. Then there is the theory of Jürgen Habermas. Finally we have the theory put forward by Cohen and Arato themselves. Up to a point all these theories incorporate the notion that public problems are caused by the rationalization process gone awry. Another similarity is that in all these three theories 'differentiation' and 'democratization' constitute key concepts. The theories differ, however, in their appreciation of these processes.

In order to shed as much light as possible on the specific therapy proposed by Cohen and Arato, I will now set about presenting these differences briefly. I will begin by comparing the theories' positions on the desirability of differentiation, before going on to compare their various views on democratization. Habermas considers social differentiation to be an irreversible process which has two sides. On the one hand, differentiation is undesirable because it gives rise to the system that is now threatening the lifeworld. On the other hand, differentiation also has considerable advantages. The rationalization of the market brought great material prosperity, while the rationalization of the lifeworld allowed man to escape the clutches of superstition, irrationality, lack of freedom and so on. Accordingly, Habermas does not advocate straightforward *de*differentiation. In fact he takes a step in the opposite direction: the only remedy Habermas comes up with in the end is even greater differentiation. He seems to regard this as the only way to salvage at least something of the ideal of the Enlightenment. Civil society can only be saved from colonialization by means of a far-reaching detachment from market and state. From this point on I will refer to this goal as the pursuit of absolute detachment. It can also be found in the work of Claus Offe (1979), John Keane (1984) and André Gorz (1994). Cohen and Arato criticize this drive towards absolute detachment, which they see as unrealistic or even naive, given the collective dimension of modern society.

Classical radical democrats are diametrically opposed to Habermas when it comes to the desirability of far-reaching differentiation: as they see it, civil society and state should be further intertwined. In other words, radical democrats are after *de*differentiation or fusion. Cohen and Arato also question the correctness of this position. Like Habermas, Cohen and Arato point to the value of differentiation for the development of society.

> 'Indeed, the ideal of free voluntary association, democratically structured and communicatively coordinated, has always informed the utopia of civil (political) society, from Aristotle to the young Marx in 1843. But such a "democratic" utopia, if totally generalized, threatens the differentiation of society that forms the basis of modernity. Moreover, from a normative point of view, any project of dedifferentiation is contradictory, because it would involve such an overburdening of the democratic

process that it would discredit democracy by associating it with political disintegration
or by opening it to subversion through covert, unregulated strategic action.' (Cohen
and Arato, 1992: 451)

Another important argument that Cohen and Arato marshal against the idea of
*de*differentiation is that differentiation is a precondition for democratic and prudent
government. According to Cohen and Arato, democracy and prudence demand that
citizens be capable of rational discussion and consensus-forming. Otherwise there is
simply no possibility of reaching a sensible consensus. In its turn, rational discussion
calls for a public realm in which communicative action takes place. Such a realm is
only present in a modern, differentiated civil society (Cohen and Arato, 1992: 439;
see also Habermas, 1992: 437).

All in all it should be clear that Cohen and Arato's theory regarding the
desirability of differentiation occupies the middle ground between two extremes.
Cohen and Arato are critical of the argument for *de*differentiation but they also
attack the argument in favour of absolute detachment. Cohen and Arato stand for
moderation. A certain measure of interweaving between state and civil society is
necessary, but this fusion should remain limited. The link between state and civil
society should be carefully designed.

'*To sum up: The "utopian horizon of ... (civil society) ..." as conceived here is based*
on preserving the boundaries between the different subsystems and the lifeworld ... (Our
utopia restricts itself to the ideal of) ... communicative coordination of action to the
institutional core of ... (civil society) ... itself, in place of imposing this organizational
principle on all of society and thus dedifferentiating the steering mechanisms and
thereby society as a whole.' (Cohen and Arato, 1992: 456)

The game of positioning that accompanies the evaluation of social differentiation
repeats itself when it comes to judging the desirability of the democratization of
society. Advocates of absolute detachment, like Habermas or Gorz, have given up
on the market and the state. In their view, these fundamental institutions are beyond
democratization . In the opposite corner we have the radical democrats, who
democratize modern society to death. Cohen and Arato are exactly in the middle
between these two positions. The democratization of society is important but it also
has its limits.

Placed in this context, it is noticeable that Cohen and Arato always seem to be in
pursuit of moderation or in search of the middle way. This moderation is clearly
expressed in the concept that they themselves coin to express the uniqueness of their
approach. Cohen and Arato (1992: 451) describe their theory as a 'self-limiting
utopia'. The fight against the colonialization of civil society goes too far when it
turns into an attempt by civil society to colonialize other fundamental institutions.
However, the ideal of a self-limiting utopia is not only an admonition levelled at
enthusiastic and restless radical democrats. Cohen and Arato's self-limiting utopia
also implies that the logic of each of the other fundamental institutions must not
overreach its boundaries. The self-limiting utopia decries every form of
colonialization.

The modernization of radical democracy is Cohen and Arato's main aim.
Bringing about a synthesis between this tradition and pluralist theory represents a

subsidiary goal. The ideal of the self-limiting utopia also opens up the possibility of achieving this goal. A self-limiting utopia no longer resists rule by a minority or an elite. The position of the pluralists, who are happy to leave government to the governors, is vindicated in this respect. At the same time the ideal of the self-limiting utopia also pays its dues to the ideas held by the radical democrats. A liberal society most definitely needs an active citizenry. However, the citizens' activities are largely limited to the confines of civil society.

Civil society from a systematic perspective II

The attempt to gain an understanding of civil society from a systematic perspective was abruptly broken off in the last section with the announcement that Habermas's dichotomy 'system and lifeworld' offers no hope of progress in this regard. However, Cohen and Arato make their own attempt to systematically articulate civil society. Now that more light has been shed on the remedy offered by Cohen and Arato, their alternative systematic definition can easily be described and its value effectively estimated. In their alternative systematic search Cohen and Arato align themselves with the work of the philosopher Hegel and the sociologist Parsons. Like them, Cohen and Arato describe civil society as:

> '*a societal realm different from the state and the economy and having the following components: (1)* Plurality: *families, informal groups, and voluntary associations whose plurality and autonomy allow for a variety of forms of life; (2)* Publicity: *institutions of culture and communication; (3)* Privacy: *a domain of individual self-development and moral choice; and (4)* Legality: *structures of general laws and basic rights needed to demarcate plurality, privacy and publicity from at least the state and, tendentially, the economy.*' (Cohen and Arato, 1992: 346)

I see this as a valuable description of civil society, certainly when viewed with a political-administrative eye. In this light, plurality is indeed an important characteristic of civil society. A rational public sphere, where discussions on public issues take place, implies differences of opinion, differences of position, in short differences. Nothing smothers discussion as completely as unanimity. That publicity is important in this regard should also be self-evident.

Another valuable aspect of this description, as I see it, is the picture it paints of modern civil society as an ambivalent creature, fraught with tension. Civil society is at the same time both political and apolitical, private and public. An exclusively political, publicly oriented civil society cannot be detached enough from the administrative process to serve as a platform for rational discussion and to use its position to support, criticize, check and correct the state. Meanwhile an exclusively apolitical, private civil society cannot connect with the political-administrative process.

A final positive point about this systematic definition is that it emphasizes that civil society is a sphere created by law. Once and for all, Cohen and Arato free thinking on civil society from the romanticism that surrounds the notion of a deinstitutionalized society without the state. A civil society cannot exist if people do not have clear political, civil and social rights which can be enforced in relation to other individuals, organizations and the state. Modern civil society and the state belong together (see also Walzer, 1995).

However, we might criticize this systematic definition for being incomprehensive. The definition fails to specify the kind of organization that might be able to play the political-administrative role that Cohen and Arato suggest within modern civil society. This systematic definition could give rise to the idea that any organization might be suited to this purpose. Judging from their discussion of social movements, this is not an opinion that Cohen and Arato appear to hold. They identify a specific attitude and a specific internal structure for organizations able to play a part in initiating and continuing rational discussions on political-administrative issues. In terms of internal structure, these organizations should be democratic in design. Otherwise, Cohen and Arato argue, they are in no position to articulate the wishes and opinions that exist within society. In terms of attitude, these organizations should take the public sphere and its internal logic seriously: they should genuinely participate in the discussion, which means being prepared to revise their opinions when sound arguments compel them to do so. In line with this last characteristic, Cohen and Arato refer to these organizations as self-reflexive. Groups which purely represent the interests of certain lobbies do not meet these criteria, since they are not prepared to expose their opinions and the interests associated with them to discussion. Organizations from 'new social movements' such as the ecological movement or the women's movement do meet these criteria according to Cohen and Arato, an opinion that to my mind says more about Cohen and Arato's political loyalties than it does about reality.

4.5 AN EVALUATION OF CIVIL SOCIETY THEORY

The evaluation of the administrative theory in the last chapter has left us with a rather disconcerting conclusion. While the administrative theorists made it clear that the indirect responsibility model is worn out and that a new mental model of social organization is needed, their attempts at designing such a new mental model failed to get off the ground. It was for this reason that I modelled my analysis of political philosophy around the question of whether the political philosophers are able to complement the thinking on co-management and thereby strengthen the direct responsibility model. This question can now be answered. For the sake of recollection, I will first briefly sum up the main areas of wear and tear in the indirect responsibility model, before going on to present an equally brief summary of the pros and cons of co-management.

The first reason why the indirect responsibility model is becoming less and less adequate is that, under modern conditions, the state will be soon overloaded in an order structured according to this model. Within such an order, the state is given almost exclusive responsibility for public issues. This is too great a responsibility, especially given the highly collective dimension of modern-day social processes.

The second reason for the downfall of the indirect responsibility model is that it makes it impossible for the state to deliver on its promise to act with prudence. The state has lost sight of the balance between the strategic and the idealistic aspects of government. Dependency on the market and the internal dynamic of power processes are the main reasons for this.

The third reason for the indirect responsibility model's loss of function is that, under modern circumstances, it is difficult to give meaning to the idea of democracy interpreted as sovereignty of the people. Given the complexity of administrative operations, the influence that citizens can exert on politics is small. It is perhaps even more painful to note that the influence of politics on the civil service also leaves a lot to be desired.

The last and most certainly not the least important reason for the decline is that, under modern conditions, societies structured according to the model of indirect responsibility will probably face a double stalemate when it comes to controlling the market. The rival tendencies with regard to planning and the power of the state can no longer be appeased.

Co-management is presented here as a possible interpretation of the direct responsibility model. Co-management counters the overburdening of the state. At the core of co-management lies the transfer of responsibility to market parties, which is clearly intended to fend off this problem. Nevertheless it must be noted that the advocates of co-management do not present this transfer of responsibility elegantly and thoroughly. Accordingly the design produced by co-management's protagonists has failed to strike a healthy balance between the need for new structures in the market and the need to safeguard the competitive order.

Co-management is also emphatically intended to increase the prudence of government. By involving market parties in policy forming, the state can govern on the basis of better information. The careful political-administrative re-evaluation of civil society should also be seen in this light. According to the advocates of co-management, the involvement of civil society increases the chance of better decision-making. Yet a number of doubts arise regarding these suggestions for promoting prudence. To begin with, one cannot be certain that cooperating with market parties will make the state less dependent on the market. Cooperation may also increase dependencies and the state may therefore feel compelled to place even greater emphasis on the strategic component of government.

Another reservation is that the advocates of co-management are unable to make a clear link between increasing prudence and re-evaluating civil society. It is also true to say that the political-administrative re-evaluation of civil society is not well-anchored from an institutional point of view.

Where the possibility of a new reconciliation between rival liberal-democratic tendencies is concerned, co-management offers no solutions. Partly due to its corporatist leanings, co-management is more likely to represent a step in the wrong direction from a liberal-democratic viewpoint. The advocates of co-management are equally unable to cope with the issue of the dilution of democracy. Here too there is a danger that co-management will actually make matters worse. The transfer of responsibility to market parties may possibly widen the hole within democracy.

Democracy

Can the political-philosophical theory of civil society help to strengthen the design put forward by the advocates of co-management? One valuable aspect of Cohen and Arato's theory is the concept of the self-limiting utopia. This concept would appear to be a workable criterion for the internal design of the fundamental institutions and the links between them. Focusing on the relationship between civil society and state, the concept warns against both absolute detachment and dedifferentiation. Like Cohen and Arato, however, I am of the opinion that the ideal of the self-limiting utopia can be more broadly applied. Self-limitation for example is also applicable to the relationship between market and state, warning against both the corporatist tendency towards fusion and the drive towards detachment encouraged by the indirect responsibility model.

Returning to the four areas of wear and tear mentioned above, I will first consider the issue of democracy. Cohen and Arato do not specifically devote attention to the possible 'gap' in democracy resulting from co-management. However, the authors do pose the more general question of how the ideal of democracy can be invested with meaning within modern society. In part the answer they give is resolutely no-nonsense in tone. Only by revising our idea of what 'sovereignty of the people' should entail, can we give the concept institutional meaning. According to Cohen and Arato, in a modern differentiated society the literal interpretation of the ideal of government by the people has to be surrendered. Cohen and Arato give a new meaning to the concept, however, by transferring it to civil society. There sovereignty of the people can be given meaning as the pursuit of active citizenship. The most important aspect of this active citizenship is that individuals should participate in debates on public matters.

Yet Cohen and Arato are anxious not to reduce democracy to active citizenship within the confines of civil society. They see it as essential that citizens (and/or civil society) exert substantial *influence* on processes within the state and the market. Otherwise too little would remain of the ideal of sovereignty of the people. Cohen and Arato regard a sound representative system as an indispensable device for ensuring this influence. However, they are not satisfied with this ideal as it is currently realized in Western democracies. In their opinion it is essential that the market and the state should be democratized further. In this respect they stress the importance of the participatory mechanism in particular.

In order to evaluate whether these proposals regarding the democratization of civil society, market and state can strengthen the direct responsibility model, we first have to establish whether they are meaningful in themselves. Cohen and Arato's ideas about sovereignty of the people as active citizenship within civil society do harmonize well with their ideal of a self-limiting utopia. Yet there is a reservation to be expressed. Active citizenship within civil society can only replace people's sovereignty in the classical sense if civil society is able to exert considerable influence on publicly relevant processes within the state and the market. Cohen and Arato rely on democratization of market and state to secure this influence. But it is precisely in this respect, with regard to the links between the fundamental institutions, that Cohen and Arato fail to convince.

The pursuit of democratization within the market is, in my opinion, not a suggestion that is well worked out by Cohen and Arato. To begin with it is not at all clear what the authors actually mean by this but, judging by their use of terms like 'self-management', their ideas seem to be fairly far-reaching. More objectionable than their vagueness, however, is the fact that democratization within businesses does not provide an answer to public collective problems pertaining to market processes. These problems have very little to do with relations between employees and owners/managers: they are collective problems that concern the relationships between citizens, businesses and the dynamics of the market. Another major shortcoming is the friction that exists between democratization and the ideal of a self-limiting utopia. The idea that the internal rationality of the market should be preserved clashes with the notion of democratization. One might argue that it is not strictly necessary to make an exclusive choice here. Adopting an ambivalent position, however, requires that one should make it very clear where the limits of democratization lie. To my mind Cohen and Arato do not define these limits clearly enough and, in failing to do so, they invite the suspicion that they are simply sitting on the fence.

Where the proposals for the democratization of the state are concerned, I am compelled to draw a similar conclusion. The ideal of the self-limiting utopia clashes with the goal of citizens' participation in political-administrative processes. Once again it is not the ambivalence itself that is objectionable but the lack of clear boundaries. This makes it difficult to pinpoint exactly what Cohen and Arato are proposing. Another significant shortcoming regarding the democratization of the state is that, in this respect, Cohen and Arato are not very innovative in institutional terms. Bearing in mind their pursuit of a self-limiting utopia, we might reasonably expect Cohen and Arato to search for ways to influence and democratize the state which lie somewhere between the classical pluralist idea of a representative system and the radical-democratic ideal of direct participation. However, Cohen and Arato leave this terrain uncharted to a large extent. As I see it, the pursuit of a self-limiting utopia implies that this is the very terrain that we should be exploring.

This critical evaluation allows us to assess how much of Cohen and Arato's thinking about democracy is relevant to the theory of co-management. The most important aspect, as I see it, is the idea of the self-limiting utopia and related notions regarding democracy within civil society. With Cohen and Arato's help, we can side-step the problem of a 'gap' in democracy. We are free to accept that people's sovereignty in the form of a government by the people is no longer a realistic ideal under modern conditions.

But Cohen and Arato's theory does not, of course, allow us to simply accept the wholesale dilution of the old democratic ideal. If that were the case, the authors would have strayed a very long way from their radical-democratic origins. Instead their theory should be seen as addressing the question of how to combat the dilution of democracy in a new way. Cohen and Arato rely on the revitalization and democratization of civil society, and, within this, encouraging the participation of citizens in processes relevant to political-administrative matters. Such a proposal seems to me to be highly compatible with the ideas of the advocates of co-management, which makes it even more of a pity that Cohen and Arato's theory is so clearly inadequate when it comes to the links between civil society, market and state.

Prudence

When it comes to prudence, the political philosophers, in my opinion, provide sound reinforcement for the theory proposed by the advocates of co-management. We have already seen how the advocates of co-management cast a cautious eye in the direction of civil society. Cohen and Arato's theory makes the connection between the political-administrative re-evaluation of civil society and the increase in the prudence of political-administrative action seem most reasonable. Civil society is the most rational sphere. The idea that greater involvement by civil society will raise the level of prudence in the state, therefore makes good sense. In support of this claim I do not see the need to follow Habermas or Cohen and Arato in their empirically disputable claim about the dominance of communicative rationality in modern-day, rationalized civil society. I am content with the observation that communicative rationality has by far the greatest *opportunity* to develop within modern civil society.

Rational debate and opinion forming within civil society do not of course improve the prudence of the state (and/or the outcome of the political-administrative process) just like that. In order to bring this about, civil society has to be able to exert influence on the state. The influence which can emerge from discussion is not in itself enough. The factors that overwhelm the idealistic component of government within the state cannot be tamed by the persuasiveness of sound arguments alone. The state also has to become more dependent on civil society. Cohen and Arato are very much aware of this and accordingly place considerable emphasis on democratization, and participation in particular. As they see it, participation, not as an end in itself but as a means to an end, strengthens the idealistic component of government. Participation by the citizens means that a counterweight can be built up within the state, an additional force that administrators have to reckon with.

Once again I see the theorizing about the link between civil society and the state as Cohen and Arato's Achilles heel. Participation as a means clashes with the ideal of the self-limiting utopia and with the position that the rationality of the civil society develops especially well if it is kept at a distance from day-to-day administrative concerns. Meanwhile, all other objections to participation (time, expertise etc.) can once again be applied.

We can also question the casual way in which Cohen and Arato treat 'prudence' and 'democracy' as simple extensions of one another. Democratic decisions are not always prudent decisions, far from it. Majorities sometimes act in ways that are not at all sensible, while in other cases the dynamics of the democratic decision-making processes can have a significant negative influence. The reverse is also true: prudence must not be seen as a replacement for democracy. Enlightened despotism is most certainly not the same thing as democracy.

Our conclusion on prudence is therefore similar to our conclusion on democracy: Cohen and Arato make useful additions to thinking on co-management, but fall short when it comes to linking the state and civil society. An obvious question would be why they are so ill-equipped to create new openings in this respect. As I see it, part of the explanation lies in the fact that Cohen and Arato's conception of the state is too negative. Under the influence of Habermas's colonialization thesis, Cohen and Arato have a tendency to interpret the state as a purely reactive organization that serves the market. In their eyes the state is all dried up normatively speaking: it acts purely on the basis of strategic considerations and can no longer be understood as a normative actor focused on the realization of liberal-democratic objectives. Accordingly, if something good is to emerge from the state, it will have to come from civil society. Under its own steam, the state can't amount to anything, normatively speaking.

Empirically, I regard this version of events as incorrect. Normatively, the state is not utterly depleted. The federal government of the United States, for example, was partly responsible for the racial education of its Southern citizens in the second half of the 20th century. But also when it comes to sustainable development in the Netherlands, parts of the state adopt an active normative role, for example by launching awareness campaigns to get people to save energy or to sort their waste before disposing of it.

From a theoretical viewpoint the negative picture that Cohen and Arato paint of the state can be criticized for inconsistency: the vision of a normatively bereft state does not tally with the image of a state that learns from the communicative debate in civil society. Moreover, this negative conception of the state causes their theory to seize up, as it were. Because of it, Cohen and Arato lack the theoretical scope to reform the state from within. They see the state by itself as being incapable of bringing anything to bloom. Its barren land can only be washed clean by civil society. We can also say that Cohen and Arato's negative conception of the state renders them incapable of developing a normative-institutional theory of the state. Yet such a theory is of the utmost importance to the ideal of the self-limiting utopia.

Overload and stalemates

A reduction in prudence and a decline in democracy are only two of the four main causes of the downfall of the indirect responsibility model as a frame of reference. Present-day societies that are structured according to this mental model are also likely to encounter the overload of the state and face two stalemates in attempting to control the market. Cohen and Arato do not discuss these other problems, however. These omissions are quite remarkable, especially considering that they emphatically recognize the overburdening of the state as a problem. In my opinion, their silence on this issue detracts significantly from the adequacy of their design. Because they ignore so many aspects of present-day government problems, their design cannot aspire to be anything more than a partial solution.

This critique on Cohen and Arato can also be formulated differently. Cohen and Arato focus all their thinking about the organization of society on the internal structure of civil society and the relationship between civil society and the state. They have relatively little to say about controlling the market. The little that they do say makes it clear that they have a fairly 'Habermasian' view of the market. Their market appears to allow hardly any space for morality or political-administrative considerations. From this it follows almost automatically that Cohen and Arato resort to the setting of strict limiting conditions to control the market. That is to say, when it comes to controlling the market, Cohen and Arato only take a few steps back from the indirect responsibility model. Seen in this light, their theory is more of a supplement to the indirect responsibility model than an attempt to transform it. Cohen and Arato do not transfer tasks or responsibilities from the state to market parties. Instead, by revitalizing civil society, they give the state a helping hand so that it is better able to stay on 'the right path'.

4.6 LOOKING BACK AND LOOKING AHEAD

In this chapter we examined the question of whether the emergent theory of civil society provides starting points for the further development of the direct responsibility model. In this process we focused in particular on the theory put forward by Cohen and Arato. The work of these two critical disciples of Habermas is regarded as a milestone in thinking on civil society.

Cohen and Arato did not write specifically about sustainability. In their thinking, environmental problems represent an important facet of a more general analysis of the problems of modern society. The core of this societal analysis is the Habermasian notion that the rationalization of society, so characteristic of Western society since the end of the Middle Ages, has taken a wrong turn from a normative point of view. The rationalization process has not maintained an equal pace within market, state and civil society. While the state and the market have undergone extensive rationalization, civil society has lagged behind. The imperialism of state and market has been largely responsible for this state of affairs. As a result of the lack of rationalization of civil society, the rational discussion of norms and values has been pushed to one side. This discussion depends on communicative action; the form of coordinated action associated with a rationalized, modern civil society.

Classical representatives of the radical democratic tradition usually propose turning this tide by means of extensive *de*differentiation of society on the one hand and by radical democratization of the state and the market on the other hand. Cohen and Arato regard themselves as offspring of this tradition but they go on to reject these standard solutions. As they see it, such an approach denies the practical and normative advantages of differentiation.

Habermas, with whom these authors feel a strong tie, proposes isolating civil society even more from the state and the market. According to Habermas the power of the market and the state has become so great that it is impossible to break this 'system'. By isolating civil society, at least communicative action and ideals like freedom and emancipation can be saved within civil society. Cohen and Arato also reject this solution. In this age of unintended consequences they see it as an impossible task to protect civil society from the effects of the activities within the state and the market.

The solution which Cohen and Arato themselves provide occupies the middle ground between these two extremes. The authors characterize their solution as the ideal of a self -limiting utopia. A self-limiting utopia means that each fundamental institution should be organized in accordance with its own rationality to the greatest possible extent. However, this should not go too far, to the extent where there is no more room for influence from other fundamental institutions. With the ideal of a self-limiting utopia Cohen and Arato attempt to steer between the Scylla of the dedifferentiated society and the Charibdus of the over-differentiated society.

A crucial aspect of Cohen and Arato's political-administrative thinking is the position of civil society. Cohen and Arato go quite some way towards Habermas's notion that the state and the market in modern-day reality form a system that is not oriented towards prudence or democracy. These authors, however, are just a little less sombre about the situation than Habermas. As they see it, state and market can be influenced under certain conditions. The system can be steered in the right direction. The first condition for this is a well-functioning, rational civil society where normative discussion can take place about the right interpretation of liberal-democratic values and, on this basis, about desirable legislation and policy. A second condition is that civil society must be able to exert influence on the market and the state. According to Cohen and Arato this influence can be effectuated by the democratization of the market and the state.

What contribution does the work of Cohen and Arato make towards the direct responsibility model, given the point at which we found ourselves at the end of the last chapter? There we were able to conclude that thinking on co-management supplied a number of ideas for the structuring of the direct responsibility model, such as the central idea that market parties should bear a certain responsibility for public issues. Nevertheless, this school of thought fell short as a fully fledged template for the direct responsibility model. The central shortcomings were that:

* thinking on co-management does not make it clear how the market as a competitive sphere can be united with the idea of the market as a sphere in which responsibility is taken.
* thinking on co-management does not deal systematically with the rediscovery of civil society.
* thinking on co-management does not make it clear why the problem of overload will not plague this mental model of social organization under present-day circumstances.
* thinking on co-management does not offer an adequate remedy for the possible detriment to democracy resulting from the discretion offered to market parties within the direct responsibility model.
* thinking on co-management cannot demonstrate convincingly that this approach to the direct responsibility model will lead to more prudent government. Co-management could also increase the state's dependence on market parties.
* thinking on co-management only provides a one-sided solution to the problems associated with rival tendencies.

The emergent theory of civil society can provide a contribution to the further development of the direct responsibility model on a number of points. For the sake of clarity I will identify these on the basis of the above list of shortcomings.

* The theory of Cohen and Arato makes no contribution to the unifiability of competition and responsibility. The theory is not market-oriented enough for this purpose.
* The theory of Cohen and Arato does make it clear why civil society has to be a crucial concept in the further development of the direct responsibility model. Civil society is the sphere of communicative action and communicative action is essential in order to reinforce the idealistic component of government.
* The problem of overload is indicated by Cohen and Arato but is not systematically incorporated into their proposed solutions.
* Where the possible gap in democracy is concerned, the thinking of Cohen and Arato does hold relevance for the direct responsibility model. Cohen and Arato make a bold attempt to adapt the idea of democracy to the modern context. As they see it democracy does not necessarily have to mean that the citizens actually rule. In the modern context, democracy can also mean that the citizens take an active part in public debate and that society is governed under influence of rational public debate. This means that Cohen and Arato reconceptualize the normative meaning of democracy and make it 'more appropriate' to the modern context.

Purely at the level of the mental model of social organization, however, Cohen and Arato's contribution disappoints. In order to strengthen the influence of civil society on the state, all they manage to come up with is democratization. In the light of their own ideal of the self-limiting utopia, this solution has definite limitations.

* Where the problem of lack of prudence is concerned, Cohen and Arato also make a substantial contribution to the further development of the direct responsibility model. Cohen and Arato show that the prudence of the state can be increased by strengthening the influence of civil society on the other fundamental institutions. However, the question of how the influence of civil society on the market and the state can be increased is not dealt with convincingly by the authors. Once again they resort to democratization and once again it must be concluded that their heavy reliance on democratization clashes with their own ideal of the self-limiting utopia.
* With respect to the rival tendencies, Cohen and Arato provide a modest contribution. The ideal of a self-limiting utopia can be seen as an idea that points the way in a precise search for the balance between rival tendencies. However, Cohen and Arato do not go into greater detail about this notion and its relationship to control over the market.

To sum up, it can be stated that the emergent theory of civil society provides useful contributions as regards a number of issues relating to the development of the direct responsibility model. The ideal of a self-limiting utopia, for example, forms an interesting idea that points the way towards determining the relationship between state, market and civil society. It also takes into account the advantages of both differentiation and dedifferentiation. Nevertheless the theory of civil society falls short on many points which form part of our specific theme. The main reason for this is that Cohen and Arato hardly pay any explicit attention to control of the free market. As a result their work is still characterized by reflexes associated with the indirect responsibility model.

This discussion of the theory of civil society completes our tour of the three disciplines. The direct responsibility model, however, has still not been worked out in enough detail. In the next chapter I will therefore launch my own attempt to arrive at a more detailed interpretation of this mental model. In doing so I will focus on facets of the mental model of social organization which concern the relationships between the fundamental institutions.

CHAPTER 5

STATE, MARKET AND CIVIL SOCIETY
IN A NEW CONFIGURATION

5.1 INTRODUCTION

Our tour of the three academic disciplines is now complete. In the light of the problems facing the environment, we have analysed the thinking on controlling the market in neoclassical economic theory, in the administrative theory on the limits of state action and in the emerging philosophical theory on civil society. What has been gained by this analysis? I would like to sum this up in two observations. To begin with, we have gathered enough material to answer the first key research question of this study. It can actually be demonstrated that the indirect responsibility model is no longer feasible. If modern liberal democracy holds to this mental model, there is a great risk that society will not be able to cope with the public issues it is facing. Given the collective dimension of action, it is no longer possible to build a social order on the idea that parties in the market only bear a limited responsibility for public issues while the state is only in a position to control the market by means of limiting conditions.

The second key question of this study has not been answered satisfactorily. Neither the economists, nor the political philosophers, nor the administrative theorists have been able to develop an alternative, direct responsibility model. The economists and the political philosophers, to my mind, are still far too attached to the indirect responsibility model. The attempt by the administrative theorists to give substance to the new mental model neglects the liberal-democratic context and, more generally, bristles with too many question marks.

The task before us in this final chapter is therefore clear. We have to try to flesh out the direct responsibility model. It goes without saying that the insights and conclusions drawn in the previous chapters will serve as a basis for the process of substantiating this mental model. To refresh the memory, I will mention two of these conclusions briefly. The first is Cohen and Arato's insight that, although the differentiation of modern-day society is desirable and irreversible, it should not be taken to extremes. 'Colonialization' is as undesirable as dedifferentiation. As I see it, the concept of the self-limiting utopia expresses this insight very well.

The second important insight is that the controlling of the market or, in more general terms, the handling of public issues has become the business of all fundamental institutions. It is no longer the exclusive province of the state: the market and civil society have to be involved as well. The market's influence on publicly relevant processes is too powerful to allow it to be exempted from political-administrative tasks. A similar claim can be made for civil society. Adequate governance under modern circumstances demands that civil society be involved in the political-administrative process. Civil society forms a reservoir of critical rationality that can be of help to the state when it comes to taking decisions, whether such help is requested or not.

Of all the disciplines analysed in this study, the administrative theorists have made the most explicit attempt to come up with a new mental model for controlling the market. Although their theory of co-management has been shown not to hold water in the final analysis, it can still be of some use to us here. In my own efforts to give shape to the direct responsibility model, I will primarily concentrate on searching for ways to deal with the five main shortcomings of co-management. These main shortcomings can be summarized as follows.

* *Competitiveness*

The advocates of co-management are not sufficiently thorough in examining the relation between the desirability of competitiveness and the desirability of responsibility in the market.

Overload

The advocates of co-management do not make it clear how and why the overburdening of the state will diminish under co-management.

Democracy

The advocates of co-management do not have a satisfactory solution to the problem that the transfer of responsibility for public issues to market parties may end up working to the detriment of the democratic identity of modern Western society.

* *Prudence*

There is a danger that the idea of co-management could have a negative influence on the state's capacity to strike a balance between the strategic and the idealistic component of government.

* *Rival tendencies in the liberal-democratic tradition*

The advocates of co-management do not realize fully enough that the liberal-democratic tradition is beset by rival tendencies. They use a one-sided solution to tackle the tension that exists between planning and non-planning.

One should not expect final answers here though. I have set myself a more modest task. A mental model of social organization contains a vision of the internal structure of each fundamental institution, along with a vision of their mutual relations. In this chapter I will concentrate primarily on the *relations* between the fundamental institutions, without devoting much attention to the internal structure of the various fundamental institutions. An important reason for this limitation is that it makes sense not to be too ambitious in our design of the new mental model. In addition the issue of the links between state, market and civil society remains underexposed in many analyses. The work of Cohen and Arato and that of many administrative theorists can serve as an example in this respect.

Focusing on the relations between the fundamental institutions makes it possible to structure this chapter in a relatively simple fashion. In each of the following three sections I will discuss one link. In Section 5.2 I will consider the relation between civil society and the state, in Section 5.3 I will analyse the relation between state and market and in Section 5.4 I will focus on the relation between the market and civil society. Final conclusions will be drawn in Section 5.5.

5.2 THE LINK BETWEEN STATE AND CIVIL SOCIETY

In the old, indirect responsibility model, the idea of differentiation is uppermost. One consequence of this is that the links between the fundamental institutions are kept to a minimum and, insofar as they do exist, they are subject to strict conditions. Where the link between state and civil society is concerned, the indirect responsibility model limits it to the periodic election of political representatives by the citizens. The aim of this link is to safeguard the democratic nature of society and the level of prudence in political-administrative action.

In this study we have encountered various authors who are of the opinion that democracy and prudence are insufficiently safeguarded in this way. These objectors include representatives of the radical-democratic tradition and a number of environmental philosophers. As they see it, state action in society lacks prudence and the representative system gives too meagre an interpretation of the idea of people's sovereignty. In order to bring about change in this respect, they propose the far-reaching democratization of society. They regard the dedifferentiation which results from this process as desirable.

The work of Cohen and Arato deals to a large extent with the links between state and civil society. In this regard it is interesting to note that Cohen and Arato resist both the indirect responsibility model and the radical-democratic tradition. They express their ideal in the concept of the self-limiting utopia. This concept enables Cohen and Arato on the one hand to articulate the idea that the differentiation of society is too valuable to abandon, while on the other hand it allows them to criticize the indirect responsibility model for its lack of democracy and prudence.

In this section we pose the question of what form the link between state and civil society should take within the direct responsibility model. In doing so I will expand on a number of important conclusions from the fourth chapter. The first is that, in principle, the concept of a 'self-limiting utopia' is a step in the right direction. The second is that Cohen and Arato are not consistent enough in designing their model. Ultimately, they too place limitless trust in democratization. The third point is that Cohen and Arato pay no heed to the possible conflicts between the goal of increasing prudence and the objective of reinforcing democracy.

In the light of these conclusions, I am able to reformulate the question of how to give shape to the direct responsibility model, and instead ask the question of whether it is possible to deal with the shortcomings in the theory put forward by Cohen and Arato. In this section I will proceed on the basis of this reinterpretation. The question regarding the link between state and civil society within the direct responsibility model will be reinterpreted as the question of how to deal with the shortcomings in the thinking of Cohen and Arato. In this section, I will concentrate mainly on the problem of the decline in prudence. I will save the problem of evaporating democracy until the section which addresses the link between market and civil society. The reason for this is fairly fundamental. The direct responsibility model lays part of the responsibility for public issues at the door of the market parties. As I see it, this means that part of the problem of a shortfall in democracy comes to lie with the market. Reinforcing the ideal of democracy therefore calls for reflection on the link between market and civil society.

Legitimacy through distance

In this study, the problem of limited prudence in state action has, for the sake of clarity been reduced to the problem that, under present-day circumstances, the idealistic component of government has become snowed under by the strategic component. In order to increase prudence in this sense, Cohen and Arato opt for a remedy that consists mainly of democratization. The underlying idea is that, through democratization, the state will be influenced by the communicative rationality that reigns in civil society. According to Cohen and Arato this will be particularly effective if democratization takes the form of participation by the citizenry in the political-administrative processes of the state. Cohen and Arato therefore treat prudence and democratization as extensions of one another.

In Section 3.2 of this study, a distinction is made between democracy as sovereignty of the people and democracy as consultation. In their work, Cohen and Arato do not make it clear which version of democratization they have opted for. Are they dealing with democratization as people's sovereignty, democratization as consultation or democratization in both senses? In the light of their ideal of the 'self limiting utopia', however, it can be deduced that, in principle, they take democratization to be a reinforcement of the opportunities for consultation. This is how I interpret their plea for democratization and as such I do not accept their reasoning. That is to say, I do not agree with the idea that prudence and the enlargement of the possibilities for consultation always serve as extensions of each other. First of all we have to ask ourselves whether the democratic process in a fully rational civil society will always lead to prudent decisions. Democratic decisions are not necessarily prudent decisions, if for no other reason than that the decision-making procedure can in itself have irrational consequences. In addition to this, the citizens in a fully rationalized civil society are confronted with situations in which their rational self-interest clashes with their collective rationality and/or their normative beliefs. Since it would be unrealistic to operate on the assumption that these citizens are moral heroes, the possibility of them making non-prudent decisions cannot be ruled out.

Secondly, civil societies as they actually exist can never be fully rationalized. In the real world, therefore, blind faith in the outcome of a democratic process is not a good idea. The third and possibly the most important argument is that, in Cohen and Arato's thinking, civil society does not derive its rationality from the individual rationality of the citizens, but from the fact that the citizens are forced into rationality in the public domain. The rationality that characterizes civil society does not occur spontaneously and is not necessarily inherent in the citizens who make up civil society. Civil society becomes rational because only good arguments count in the public sphere. Citizens are compelled to act rationally. The idea that the state can become more rational as a result of democratization ignores this fact. It equates the rationality of civil society with the rationality of individual citizens. Democratization, after all, does not bring civil society into the state. It doesn't even bring the citizens into the state. Democratization always comes about by virtue of the fact that a number of specific citizens acquire influence within the state. On the basis of experience, there is no reason to assume that these citizens stand for prudent opinions.

In other words, democratization, especially in the form of participation, is the interweaving of the state with parties from civil society. In this it resembles corporatism, which ties the state to specific market parties. If the ideal of the self-limiting utopia pledges resistance to corporatism then, at the very least, caution should be exercised when it comes to the pursuit of democratization (in the form of participation).

If these considerations lead us to pay serious heed to the ideal of a self-limiting utopia, then, to my mind, it follows that prudent government calls for a certain *distance* to be maintained between the state and specific parties from civil society. In adopting this position it is not of course my intention to flatly contradict Cohen and Arato. It would be nonsense to claim that the prudence of the state enjoys no benefit at all from democratic processes. The point I would like to make is that possible friction between democracy and prudence easily escapes notice when the pursuit of prudence is spuriously equated with the pursuit of democracy. A prudent state requires a degree of autonomy.

In my view, this position follows from accepting the idea that modern society is differentiated. Differentiation makes a specialism of government. It gives each fundamental institution specific capabilities, responsibilities and roles. The price to be paid for this is that each fundamental institution is also denied certain capacities as a result. For example, to a large extent differentiation denies the state the opportunity to act efficiently. This is simply the down side of the weighty collective responsibilities and extensive powers accorded to the state in modern society. As a result, the actions by and within the state must meet high demands in terms of such aspects as lawfulness and rationality (see also Durkheim, 1922). This makes the state by definition an organization that is expensive and slow to act.

By the same token, functional differentiation relieves the individuals in civil society of administrative tasks and in return gives them greater freedom to act in other areas. The price they pay for this is that limits are set on the individual's ability to keep abreast of administrative processes. The perspective, time and expertise needed in order to make administrative decisions become subject to an increasing amount of pressure. In the actual political-administrative process, the ordinary citizen only has a walk-on part.

A good example of this is the car as a means of transport. From the perspective of the individual citizen, the car is a more or less necessary part of keeping one's life up and running in the modern world. Bearing this in mind, it may even be unreasonable to demand an enthusiastic response from individual citizens when it comes to the introduction of measures to restrict the use of cars. From an administrative perspective, however, the car must be seen as a major cause of environmental problems such as acid rain and fragmentation of environmental areas. A state that wants to make a serious attempt at achieving sustainability therefore cannot rule out unpopular measures against road traffic.

When it comes to a goal like sustainability, it is particularly important to underline the need for distance between the state and individual citizens within civil society. Sustainability requires a state that keeps its eye on long-term considerations and that is prepared to take responsible choices. There is little chance that such a state will receive enough direct and immediate support for concrete measures from a modern civil society in which the citizens are neither disposed nor obliged to take administrative responsibility. In such a situation we cannot and should not expect too much prudence from the citizens. They are too far removed from the day-to-day process of administration for that. This distance makes it difficult to obtain the facts and appreciate their value.

It is also true to say that, under modern-day circumstances, a highly democratic government would demand too much in the way of sacrifice from its citizens. In practice, such heroism is not to be found in great quantities among ordinary people. We see ample illustration of this when, for example, citizens have to choose between their own employment opportunities and sustainability. Many then opt for the former. The campaign by lumberjacks in the north-west United States using the slogan 'Save a job, kill an owl' provides striking evidence of this point. As I see it, campaigns of this kind do not prove that people regard sustainability as unimportant. It is primarily evidence of the need for an autonomous state in a differentiated society.

A Dutch example of the clash between democracy and sustainability centres on the policy regarding the placing of wind turbines. The aim of this policy is to increase the proportion of the Netherlands' energy needs supplied by wind power. At the same time the policy is supposed to serve as an example of the feasibility of grassroots democracy in modern-day society. The result is that, in comparison with other infrastructure projects, the placing of a windmill involves a disproportionate amount of effort. Windmills are considered ugly and noisy, so no one wants a windmill in their back yard.

Does this mean that citizens are irresponsible? Will my argument eventually bring me to this conservative hobbyhorse? Not necessarily. I am not accusing citizens of acting irresponsibly in the public sphere. The point I am making is that reasonable limits should be set on the matters for which citizens can be held directly responsible in modern society. My plea for a certain measure of autonomy for the state is not meant as a direct contribution to the debate on the responsibility or irresponsibility of citizens. I would much rather see it as part of the debate on the institutional organization of modern democracy. As such, my argument is aimed against the wholehearted pursuit of democratization at the expense of all else.

I also regard my argument as a contribution to the debate on the way in which administrators and politicians should *represent* the voters (citizens). In the debate on representation within modern democracy, two opinions stand opposite one another. According to the 'parochial idea' representation means the direct defence of a person's interests (see Bailyn, 1982: 161-175). Supporters of this idea feel that a representative ought to bow to the wishes of her voters and what they view as being in their own best interests in a certain time and place. The parochial representative is a direct replacement, someone who reflects the opinions of those she represents as faithfully as possible.

Edmund Burke is exemplary of the other vision. He can go along with the parochial view to the extent that

> '*their wishes* (the wishes of those who are represented) *ought to have great weight with him* (the representative); *their opinions high respect; their business his unremitted attention. It is his duty to sacrifice his repose, his pleasure, his satisfactions, to theirs, - and above all, ever, and in all cases, to prefer their interests to his own'*. (Burke in Ankersmit, 1997: 12).

But Burke rejects the idea that the representative should *identify* with the represented. Representative and represented must not be allowed to forget that they occupy different roles in a differentiated society. As such they view issues from a different perspective and the likelihood is great that they will also hold differing opinions. The Burkian vision of proper representation is therefore that the representative should take the views and the interests of those she represents into serious consideration when forming a judgement on a public issue, but that she should ultimately form her own opinion from within her role as a public figure and from the perspective of the common good. In Burke's own words:

> '*Parliament is not a congress of ambassadors from different and hostile interests, which interests each must maintain, as an agent and advocate, against other agents and advocates; but parliament is a* deliberative *assembly of one nation, with* one *interest, that of the whole - where not local prejudices, ought to guide, but the general good, resulting from the general reason of the whole. You choose a member, indeed; but when you have chosen him, he is not a member of Bristol, but he is a member of* Parliament'. (idem, 13)

My position that the state in modern society should have a degree of autonomy, is an argument for a Burkian form of representation. It is not necessarily an argument against democratization in any form. It is an argument in favour of the recognition that a legitimate administration can and indeed should also maintain a distance from its citizens. In this sense it is an argument against limitless democratization.

There would appear to be something of an inconsistency in my argument. On the one hand I am making a plea for a political-administrative revitalization of civil society. My main argument in favour of this is that the government of society could profit from the rationality which this sphere has to offer. On the other hand I am arguing for a certain distance between the state and the citizens who are part of civil society. My main argument for this is that the individual citizens are not necessarily prudent, indeed their representation may not even extend beyond their own interests.

As I see it, there is no head-on conflict between these opinions, however. In the first place, 'a certain distance' is not the same thing as alienation. The distance I am advocating refers only to the need to give the state a certain amount of space. It is not an argument against all forms of democratization. Secondly, I have argued that the rationality of civil society is not a function of the rationality of individual citizens but a function of the structure of the public domain that forces people to be rational. Making the state more prudent therefore does not necessarily imply democratization.

In this section I have set myself the task of outlining the new direct responsibility model, in particular with regard to the link between civil society and the state. I would now like to present my first suggestion. If prudence requires autonomy, and autonomy flourishes under a Burkian form of representation, then the question becomes: how can Burkian representation be reinforced?

In my opinion, this involves improving and extending the design of the safeguards against the might of the state. Present-day safeguards which keep the actions of the state in line, are either procedural (e.g. the demand of lawfulness) or they set the limits of the sphere within which the state can act (e.g. the distinction between private and public). The *subjects* or *themes* which the state is allowed to deal with are relatively unlimited. For example, the Dutch constitution details what the state can, may or should be concerned with. Little is said about the issues in which the state *is not allowed* to intervene. I think it is possible to support the Burkian representation by drawing boundaries in this respect as well. This would mean that some subjects may no longer feature on the political-administrative agenda, while by the same token others might be given more of a chance.

If limiting the core activities is to reinforce the Burkian representation, then of course these core activities must primarily include those liberal-democratic ideals that are difficult to realize; the ideals whereby the weaker, the less resilient, the less powerful or the long-term are at stake. The ideals which already enjoy enough support from society, like economic prosperity, should be excluded. The Dutch fisheries sector has already served as an example several times in this study. Establishing core activities in the fisheries sector would mean for example that the Dutch government should place more emphasis on social issues (the problem of underdeveloped areas), on the biological problem of overfishing and on other ecological issues. The economic responsibility that the state now takes for the fisheries sector would fall away to a significant extent. Of course this would not apply to those economic tasks which the market does not always regulate very well itself, such as innovation, education and securing stability.

In my opinion this argument in favour of thematically restricting the tasks of the state forms a good supplement to the thinking of the Dutch political philosopher Frank Ankersmit. Ankersmit (1995) is one of the few authors who, like me, argues in favour of extending the autonomy of the state. According to Ankersmit, political parties, as the core intermediary institutions between state and civil society, should take the lead in creating the necessary distance. Stated simply, Ankersmit argues that political parties should distance themselves more from citizens' self-interest in their party manifestos, by presenting themselves less as 'catch all' parties. Instead they should be clearer about the public ideals that they stand for. Ankersmit (1995: 222) therefore argues in favour of 'ideologizing'. As he sees it, a party which acts on the basis of a detailed ideology (about the good society) positions itself at exactly the right distance to the citizenry.

The task that Ankersmit assigns to the political parties is a meaningful one. Yet it places a heavy responsibility on these vulnerable political-administrative institutions. Individual political parties cannot simply convert to a Burkian type of representation all by themselves. The risk is great that a party that makes such a move will disappear from the electoral stage before long. The extension of safeguards may well be a way to escape this trap: specifying the tasks of the state makes it possible for parties to take their responsibilities by forcing all parties to do so.

Prudence by means of functional representation

So far I have stated that although Cohen and Arato are right in arguing that the prudence of the state should be increased by strengthening the influence of civil society on the state, their flirt with democratization should be evaluated critically. In accordance with this view, my second suggestion in outlining the direct responsibility model is that the influence of civil society on the state should be increased but without retreating into democratization (as consultation). The idea is to increase the accountability of the state by means of functional representation.

Arguments for increasing accountability (Freeman III and Kneese, 1973; Paehlke, 1989) usually culminate in a heartfelt appeal to make government more open. As I see it, openness is not a sufficient guarantee of better administrative accountability. Increasing accountability also requires that the citizens be capable of assessing the true worth of the state's arguments. A sound relationship based on accountability only works if there is also the possibility of effective monitoring.

This is exactly what is lacking in modern society. Monitoring is a difficult task. The complexity of the political-administrative process makes it difficult for the citizens, as audience, to interpret the performance. In other words, the complexity of administrative issues creates in effect a watershed between experts and laymen. For the layman, lack of time alone makes it almost impossible to obtain a penetrating insight into political-administrative processes. Openness of government therefore offers an insufficient guarantee of effective monitoring. The facts presented and the way in which they are brought into relation with one another can hardly be verified at all.

There are many ways of illustrating this point. For example, Dutch environmental expert Lucas Reijnders (1993: 20) writes that '*CO2 emissions ... (in the Netherlands in the period 1990-1994) show a clear upward trend. Corrected to take into account conjunctural and temperature fluctuations, the increase is around 1 per cent per year.*' Who is capable of verifying this statement? In order to do so, it would take an environmental expert, a meteorologist, a chemist, an economist, a statistician, philosopher of science and someone with in-depth practical knowledge. Increasing accountability under modern circumstances therefore requires new and special institutions. Administrative experts need well-equipped monitors at their side, from and for the benefit of civil society. These experts should be given the task of monitoring the political-administrative process and passing the relevant information on to the citizens in civil society. They should also deliver an initial interpretation of this information to the citizens.

Within liberal democracy, the people's representatives form the most important monitoring institution in the direct service of the citizens. This institution suffers from a lack of time and manpower and is therefore not always able to acquit itself of this monitoring task properly. Equipping parliament more extensively, for example by increasing the number of staff, would seem to be an obvious solution.

Yet, as I see it, this is not the right way to increase the supervision of the state. Various authors, like Ankersmit, Cohen and Arato, point out that the role and position of parliament in modern society is changing slowly but surely. The people's representatives were once intended to be civil society's outpost within the state, or - to use my own terms - the guardian of the idealistic component of government within this fundamental institution. Parliament is now losing this position. The people's representatives are changing from a bridge between civil society and state into a 'proto-state', an institution which has become part of the state, and which is certainly every bit as sensitive to the strategic component of government as the rest of the state.

The introduction of competition is often a good way of getting things to work more effectively. In relation to the state, Van den Doel (1978) has developed this idea in a radical direction by proposing that every administrative task should be carried out by two competing administrative bodies. As a general proposal this does not seem very realistic. Nevertheless, a more specific application of this idea by doubling the monitoring function of parliament might not such a bad move. The representation of civil society in the state should be reinforced. In addition to parliament, a second body is needed to represent civil society within the state.

The first task of this body would be to keep an eye on the doings and the dealings of the state. Just as important is the task of keeping civil society informed about the political-administrative process and of facilitating the exchange between debates going on within civil society and the state. This new body should develop a tense and ambivalent relationship with parliament. On the one hand parliament should of course be an ally to the new body, given that both institutions are there to keep an eye on the government and public services on behalf of the citizens. On the other hand the new body also has to keep its distance from parliament. After all, parliament is turning into proto-state. The new body therefore has to keep an eye on the people's representatives as well.

Given that expertise would be of great importance within such a body, functional representation seems like a logical next step. Functional representation is a form of representation in which the citizens allow themselves to be represented by selected organizations which fulfil a specific position or task within civil society. When it comes to improving the sustainability of the market process, what springs to mind is a balance between nature conservation and environmental organizations on the one hand and market organizations on the other hand. Periodic selection of these organizations could be carried out by a parliamentary council. Expertise, credibility and support from the market or civil society should be important selection criteria.

Given the primacy of the monitoring task, it would not be useful for this new body to be related to the state in the same way as parliament and the state are related: that is to say as a separate body in session which is linked to the policy process by virtue of certain powers and rights. I think it would be more effective if membership of the new body gave the right to observe the policy process in a participatory way. In addition, a place in the new body should involve the right to be heard as part of the policy process and offer the possibility of informing or advising the government and public services either spontaneously or on request. In this way, the state can also benefit from the intervention of these 'busybodies'.

My proposal does not appear to contain much that is new when seen from a practical perspective. In the United States, but also in the Netherlands, it is common administrative practice for interested parties to be consulted in almost all phases of the policy process. Yet I am not putting forward this proposal in order to legitimize a situation which has already developed in practice and I diverge from this approach on three elementary points. First of all, in my proposal participation by interest groups in the policy process is no longer at the discretion of the state. Instead I give such participation the status of a formally defined right. This difference is an important one. We see that the state has a tendency to shut out (specific) interest groups, and in particular the more 'troublesome elements' from civil society, in accordance with its own agenda, for example when there is the threat of a conflict with market parties (see for example Van Vliet, 1992).

Secondly consultation within the existing situation often tends to serve the strategic component of government. Consultation is used to reach compromises. In my proposal, consultation is intended to inform civil society, and in doing so to strengthen the idealistic component of government.

Thirdly the task of the functional representatives in my proposal is centred emphatically on monitoring. The gives functional representation a whole new meaning: it is no longer an instrument of the state, but of civil society.

Whoever argues in favour of greater autonomy on the part of the state invites the suspicion of being an authoritarian elitist: someone who, like William Ophuls, thinks that the average citizen will never willingly opt for sustainability and who therefore attaches restrictions to democracy, albeit temporary ones, for the sake of future generations. Since I want to remain within the liberal-democratic tradition, it is important that I explicitly absolve myself from any such suspicion. To this end, I would like to use the last part of this section to set out a number of important differences between my position and that of the authoritarian elitist.

To begin with, many authoritarian thinkers argue that the state should be unfettered. The authoritarian elitists experience the safeguards against the power of the state as a hindrance which is no longer responsible in the light of latter-day public problems. If the state were given greater power, it would be able to act more effectively. I think that history has shown that the state does not deserve such a degree of trust. Whoever does away with these safeguards, exposes society to enormous risks, from arbitrariness to nepotism, and from partisanship to patronage. What is more, from a historical point of view it is certainly not the case that the state has always regarded protection of those unable to protect themselves as its natural task. Intervention by the state has always led to more of the natural world being destroyed than protected. I am therefore not arguing in favour of 'decisive state action' above all else. If anything I am arguing for the opposite: I would like to see greater safeguards surrounding the action of the state. By taking away the state's opportunity to represent certain interests, it may be hoped that other interests will be given a chance.

Another difference with the authoritarian-elitists is that they seek to put the blame for environmental problems largely in the lap of the citizens themselves. I, on the other hand, point to inadequate control of the market as the main cause. The elitists also tend towards keeping democracy as such on ice, for example by fiddling around with the electoral system. Instead, I would argue in favour of a difficult, ambivalent and tense double-bind relationship between civil society and state. On the one hand differentiation calls for distance between these two fundamental institutions. The state does need a certain room for manoeuvre. On the other hand it is abundantly clear that democracy is essential and desirable for both strategic and normative reasons.

Another difference relates to the fact that, sooner or later, authoritarian-elitists come up with the notion that people can be divided into types: those capable of governing thanks to their birthright or innate capacities, and those who in principle are simply not capable of doing so. My argument does not rest on such elitism, nor does it have to. If there is any strain of elitism in my argument, then it is a form of 'role elitism'. People should be given a responsibility in accordance with their role. I do not therefore concentrate on the typically elitist question of how to ensure that the right people end up in the right places (Ankersmit, 1997: 11; Pels, 1993). In my own institutional design, I am more interested in the question of how we can encourage and compel individuals and organizations to take the responsibility that corresponds to their role.

5.3 THE LINK BETWEEN STATE AND MARKET

The traditional, indirect responsibility model argues for the strict separation of state and market. The influence of the state on market processes should be minimalized. The same is true of the possible influence of market parties on the state. With the help of administrative theory, I have made it clear that the indirect responsibility model's vision of the link between the state and the market is no longer feasible in the modern context. It gives rise to a number of problems. To begin with, the strict division of tasks and responsibilities leads to an overburdening of the state. In addition to this, it is becoming increasingly difficult for the state to control the market prudently. The indirect responsibility model's assumption that the state has the potential to keep a firm grip on the market turns out to be a fiction. The last and certainly not the least important problem with the old mental models' vision is that, under modern circumstances, it fails to provide a workable balance between the rival tendencies within liberal democracy. The pursuit of planning and the pursuit of non-planning can no longer be simultaneously appeased.

The administrative theorists have tried to come up with some alternatives to the indirect responsibility model. Co-management was considered to be the best of these proposals. One very important notion in co-management is that public responsibility has to be redistributed. The market must be given some responsibility for public issues. In addition to these considerations, the advocates of co-management feel that market parties and the state should be much more emphatically tied together, for example by establishing regular consultative bodies where state representatives and market parties meet each other. These ties are intended to facilitate better cooperation between the state and market parties. Partly in the interests of this improved cooperation, advocates of co-management also argue for a transformation in the policy armamentarium, with hierarchical-juridical instruments making way for communicative, non-juridical instruments.

Co-management has a number of unmistakable advantages. The redistribution of responsibility tackles the root cause of overburdening. Yet the idea of co-management also has a number of design flaws. For example the strategy of reducing the dependency of the state by means of cooperation with market parties, is definitely not without its dangers. There is a real risk of making the state even more dependent on the market. Co-management also pays little attention to the risks presented by a corporatist system. Two other limitations are the readiness of the advocates of co-management to ignore the dangers of dejuridification and their tendency to overlook the fact that it is no straightforward matter for a market party to take on wider responsibilities.

In this section I want to outline the alternative, direct responsibility model with regard to the link between state and market. As in the previous section, I will do so by building on conclusions already arrived at. Accordingly I will present my ideas about the direct responsibility model as amendments to the idea of co-management. I have formulated seven of these amendments. Generally speaking, what I am looking for in the direct responsibility model is a middle path between co-management and the indirect responsibility model. In my opinion a great many of co-management's shortcomings can be said to represent too radical a departure from the indirect responsibility model. Just like Cohen and Arato, the advocates of co-management are thinking on the rebound. They take too little account of the need for distance between the state and market. Where the indirect responsibility model is exclusively concerned with separation and differentiation, the advocates of co-management focus solely on cooperation and interweaving. My amendments are geared towards reintroducing the concept of differentiation into the proposals of co-management, allowing us to preserve some of the sensible aspects of the old mental model in the new.

My first amendment is that the various levels of analysis should be more explicitly and more emphatically separated in the theory of co-management. When shaping the relationship between the state and the market we have to think about (a) redistributing tasks and responsibilities, (b) organizing interaction, and (c) the measures available for exerting control. Co-management's advocates fail to adequately distinguish between these levels of analysis, in effect reducing them to extensions of one another. For example, co-management theory conflates the idea that public responsibility should be redistributed both with the idea that corporatist structures are desirable and with the idea that hierarchical government is outmoded. In my opinion, there is no logical relationship between these various ideas. Protestant theories of the market system, for example, have always rejected corporatism, yet readily embraced the idea that market parties should bear some responsibility for public issues (Couwenberg, 1953).

The second amendment specifically concerns the redistribution of tasks and responsibilities. Advocates of co-management assign some public responsibility to market parties but are not very specific about doing so. I think that tasks and responsibilities ought to be mapped out in considerable detail in the direct responsibility model. Tasks and responsibilities should be as clearly, meticulously and *exclusively* allocated either to market parties or to the state, as they are in the indirect responsibility model. Undivided and collective responsibilities are a direct invitation to creeping corporatist leanings. The danger is then great that tasks will be neglected because no single body considers itself responsible or indeed can be held responsible.

My third amendment is that the redistribution of tasks and responsibilities should take into account the consequences of the differentiation of society. To start with, this means that the final responsibility for public issues should remain with the state. A possible transfer of tasks and responsibilities does not relieve the state of its general corrective role. Taking into account the consequences of the differentiation of society also implies that the transfer of public responsibilities to market parties ought to be accompanied by stiff measures to incite market parties to act responsibly. To approach this issue from a different perspective, if market parties are required to act responsibly, then markets should be organized in such a way that the responsible party is rewarded as generously as possible for such conduct. Responsibility implies freedom but this should not imply that we are not allowed to support market parties in taking their responsibility. In some markets it might be enough to create transparency while facilitating the role of the NGOs who closely examine the conduct of market parties. In other markets, more extensive measures or other alternatives might be needed. For example in markets where stiff competition reigns, it might be necessary to back up the transfer of public responsibility with a solid sanction policy clearly established in advance.

A good present-day example of co-management which takes into account the rationality of the market and the need to lay the final responsibility firmly at the door of the state, can be found in the Netherlands' policy on cockle and mussel fishing in the Wadden Sea. The state has done more than simply make the fishermen partly responsible for the public task of ensuring that the birds in the area have enough to eat. It has also announced a plan to ban fishing in a large part of the Wadden Sea if the birds are not left with enough to eat, regardless of whether the fishermen are responsible for this shortage or not. This clearly defined sanction policy enables the fishermen, as it were, to take their own responsibility (given that in this small market there is already an infrastructure available to the fishermen that allows them to coordinate their actions). At the same time it also confirms the final responsibility of the state.

The fourth amendment relates to the organization of interaction between the state and market parties. In my opinion, the direct responsibility model demands that, at least in highly competitive situations, the market should be enriched at meso level with structures within which consultation between market parties can take place, where agreements can be made and where the new public responsibility of market parties can be translated into binding rules. To this end, these structures will have to be institutionally equipped in such a way that relevant market parties really feel the weight of their obligations. Without institutional embedding of this kind, we cannot and should not expect market parties to take on public responsibility, at least not a responsibility which goes significantly beyond that of normal market morality. This would entail far too great a risk for them.

However, such meso structures also imply a great risk to society. There is a danger that the market parties will only decide to enter into agreements which work to their collective advantage, and which end up having very little to do with taking public responsibility. In order to stave off this danger, it seems like a good idea to demand that all decisions taken within the meso structure be confirmed by a governmental institution. The task of this institution must be to examine whether the decisions taken are compatible with the pursuit of competitiveness within the market.

My fifth amendment covers the further organization of the structures at meso level. Insofar as the advocates of co-management deal with this subject, they take it more or less for granted that these structures should also involve a degree of state representation. This is not an assumption that I share. In my opinion, the state should take no part at all in these meso structures. This position is once again motivated by fear of the disadvantages of corporatism and the attendant need to clearly distinguish between state and market. For one thing, when the state is nestled in meso structures, its task does not really become any easier. It is still involved in all facets of handling the public issue in question. Besides, involvement of this kind increases the risk that the interests of the state and market parties will become so intertwined that they threaten the plural character of society. Furthermore, it poses a threat to the role of the state as a guardian of competitiveness.

My sixth amendment relates to the measures available when it comes to implementing policy. I think that the protagonists of the direct responsibility model ought to scrutinize and reconceptualize this part of the theory on co-management very carefully. In my opinion, authors such as Teubner and Willke point in the right direction when they speak of the desirability of 'reflexive law'. Reflexive law consists of legislation that helps actors to act autonomously and therefore responsibly. As such it fits in well with the direct responsibility model. However, there is a problem with the way in which this general idea is usually given concrete meaning within the theory on co-management. Advocates of co-management construct a contradiction between responsible action and stiff laws. In accordance with this they search for measures which coerce market parties as little as possible. In my opinion hierarchical measures and responsibility can easily go hand in hand. Although the responsibility that market parties are expected to take upon themselves under the direct responsibility model cannot by definition simply be forced out of them by law (in that case, actors would not be taking public responsibility but just obeying the law), stringent laws do not hinder responsible behaviour. It makes more sense to say that they are a condition for responsible behaviour. Strict laws and strict enforcement of these laws reassure well-intentioned market parties that all others parties are also subject to considerable obligations. Of course, this is not to say that such laws should be inflexible or divorced from reality. The whole point of transferring public responsibility to market parties is to overcome the traditional disadvantages of hierarchical government.

The seventh amendment has to do with democracy. The direct responsibility model fits in well with the ideal of self-rule, which is so central to the liberal-democratic tradition (Mill, 1848; 312-314; See also: De Haan, 1993). It is, however, somewhat harder to reconcile this model with the ideal of democracy. Transferring responsibility for public issues to market parties after all implies that the right to decide how to deal with these issues also becomes the province of the market party concerned. The direct responsibility model has to answer for this loss of democracy. The theory on co-management is not very helpful here since it has very little to say on the matter. As I see it the direct responsibility model can cope with the loss of democracy in three ways. First, advocates of the model can qualify the exaggerated vision of democracy that gives rise to the issue in the first place. Why should democracy mean that the citizens have the right to decide about all issues that are publicly relevant? Democracy means that they have a right to decide in the public domain. But in modern society public issues and the public domain are not one and the same thing (see Section 3.2). Second, the advocates of the direct responsibility model can use Cohen and Arato's reconceptualization of democracy to further underline the assertion that the discretion of business on public issues is not necessarily in conflict with democracy. But all this is not enough. Third, the advocates of the direct responsibility model also have to think of ways to address the loss of democracy. I will present my suggestions on this matter in the next section, in an attempt to repair any damage done by rethinking the relation between market and civil society.

5.4 THE LINK BETWEEN CIVIL SOCIETY AND THE MARKET

The link between market and civil society tends to be treated like something of a poor relation when it comes to matters of governance. The relationship between these two fundamental institutions is not the focus of serious attention within economics, political philosophy or administration as an academic discipline. This state of deprivation can be understood as the inheritance of the indirect responsibility model, which maintains that there neither is, nor should be, administratively relevant contact between these two fundamental institutions. The Berlin Wall which, administratively speaking, stands between market and civil society is seen as a desirable feature from within the indirect responsible model.

The logic behind this vision has everything to do with the conceptualization of the market within the indirect responsibility model. In a market where people are geared towards their own economic success within the confines of the law and the standards of decency, it makes no sense to bombard them with demands from civil society. Furthermore, given the fact that actors in such a market are forced to bow to the rigours of market discipline, they are hardly likely to pay much heed to civil society's demands anyway.

The neglect of the link between the market and civil society is also closely related to the indirect responsibility model's view of the state as the spider in the web of public issues. This leads to the state's respective links with the market and civil society being given far more prominence in the indirect responsibility model than the link between market and civil society.

The indirect responsibility model is bankrupt. The state cannot cope with the tasks it has been given. This implies that society can no longer permit itself the luxury of seeing the market as a sphere of limited responsibility. It also means that the link between market and civil society has to be given administrative relevance. The time has come to topple the Berlin Wall.

As Cohen and Arato (1992) rightly state, this conclusion has not yet filtered through to the world of political philosophy. Neoclassical economics also pays little attention to it. Among administrative theorists, however, interest in this area is on the rise. Albert Weale (1992), for example, is looking for ways to get representatives from civil society and market parties together around the table to discuss and negotiate the ways in which market parties should give substance to their public responsibility. Yet despite such insights, no fully crystallized theory has been developed. In general, the only suggestion put forward is that interest groups be allowed to attend corporatist consultations between market parties and the state.

The placidity of the theoretical waters is beginning to form an ever sharper contrast with the turbulence of the real world. There we see that organizations from civil society, not least environmental organizations, are far more ready to hold market parties directly accountable in terms of their social responsibilities. Such organizations pay little heed to the limited public responsibility accorded to market parties by the indirect responsibility model. As far as they are concerned, market parties most definitely do have an undeniable public responsibility. Many probably share the view expressed by Peterse (1990) and Donaldson (1982) that these responsibilities are a direct consequence of the crucial social position that market parties occupy nowadays. This is why organizations like Greenpeace, Amnesty International and the World Wide Fund for Nature (WWF) publicly seek contact with market parties, enter into conflict with them or seek out the possibilities for consultation, compromise and consensus.

Meanwhile companies increasingly feel the need to publicly explain their deeds to citizens and consumers. One example is the interview given by Wim Dik, the head of KPN Telecom, to a Dutch daily newspaper (NRC Handelsblad) following the announcement that his company had clinched a deal with the Chinese army (Giebels, 1998: 17). In the interview Mr Dik is at pains to account for his company's ties with a regime responsible for human rights violations.

There is also evidence that companies are cautiously seeking contact with organizations from civil society in order to enter into consultation about controversial issues, often behind the back of the state. A good example of this process is the U-turn Shell has made in its company policy. This formerly insular company is now prepared to take a more open approach when it comes to public responsibility. It organizes discussions with NGOs on a regular basis and draws up a social and environmental annual report, in addition to its financial one. In this context, the Marine Stewardship Council (MSC) is another interesting example. MSC is a private, non-profit organization that has been set up by a market organization (Unilever) and an organization from civil society (World Wide Fund for Nature), without the involvement of the state. The MSC is working to bring about a sustainable fisheries sector. For this purpose the MSC has introduced an eco-label for fish products and is trying to persuade as many producers and consumers as possible of its importance.

In addition to such initiatives, more and more companies are committing themselves to self-imposed codes of conduct with regard to public matters (see also Van Luijk, 1993: 190-218). In some cases, an entire business sector will adopt a code of this kind. A good example of this is the code of conduct that most producers of baby and infant milk products in the Western hemisphere introduced after they were dragged through the mud for their aggressive sales methods in developing countries.

Innovations are a sign of changing circumstances. The growing interaction between market and civil society, and the greater public responsibility being pushed in the direction of market parties can be seen as a process of political-administrative innovation, reflecting the extent to which the indirect responsibility model is losing its grip on social reality. On the one hand the citizens are beginning to feel the effects of the overload of the state. The state is forced to neglect important public problems. It often finds itself powerless or it responds inadequately, or too slowly and too weakly. On the other hand citizens are beginning to realize that, for public issues pertaining to the market, they don't always need to go through the state. The indirect responsibility model's idea that there is and ought to be a sharp division between the role of the citizen and the role of the consumer needs to be re-examined. The power that consumers enjoy in the market can be made relevant for public issues, especially when people succeed in motivating others to vote with their spending power.

Market parties in their turn are becoming aware that a straightforward appeal to limited public responsibility is not accepted as unquestioningly as it used to be. A considerable percentage of consumers take public considerations into account in at least some of their purchase decisions (e.g. free-range eggs and phosphate-free detergents). In addition to this, a degree of sensitivity to public issues can prove opportune from the point of view of a company's staff policy. A clean company reputation can figure quite prominently as a fringe benefit (Paine, 1997).

The shifts in the relationship between civil society and market that are occurring nowadays can be interpreted as attempts to escape the straightjacket of the indirect responsibility model. It is not yet apparent where this process of change will take us.

The contacts between civil society and market parties still tend to be too awkward and unpredictable to allow a clear view of such matters. There is no such thing as a set pattern in this regard. It often takes a huge amount of effort to get a serious discussion off the ground between a market party and organizations from civil society. Often discussion has to be triggered by a spirited campaign by organizations from civil society. Even then public dialogue normally only becomes possible once the companies involved realize that they can't get away with simply ignoring such a campaign, or once a counter-campaign has failed and the threat of a consumer boycott is in the air (see also Riemsdijk, 1996).

In the rest of this section I will try to further outline the direct responsibility model with regard to the link between market and civil society. My efforts will be inspired by the recent changes that have taken place in practice. Given that many recent developments fit in well with the direct responsibility model, the question one might ask is: where do things go wrong from the perspective of the direct responsibility model? What is needed to boost the possible influence of civil society on the market and to thereby make it easier for market parties to take responsibility for public issues?

As I see it, a short answer to this question is not hard to find: a new political-administrative relationship needs to be developed between civil society and the market. A new channel, as it were, needs to be carved out between these fundamental institutions. The existing relationship between producer and consumer should be supplemented by a relationship between citizen and market party. Individual citizens and small-scale individual businesses are not the most important actors when it comes to maintaining this relationship. It is probably more desirable and more practical to put this task in the hands of larger companies and representative organizations from civil society.

For the sake of a sound understanding it is important to place this idea in the context of the actual developments which are currently taking place. On the one hand my argument follows on naturally from such developments. After all we can see that market parties and organizations from civil society really are coming into contact with one another. On the other hand my proposal should be seen in opposition to actual developments. The existing political-administrative contact between civil society and market parties hitches a ride on the back of the consumer-producer relationship that exists between the market and civil society. As such, this political-administrative contact does not have an independent status. Although this 'parasitic' contact is of course very important, I don't think it could ever form an adequate basis for the development of a political-administrative relationship between market and civil society. At some point the boundaries of consumer sovereignty are reached or the obligations placed upon market parties become excessive.

But why do we need a new political-administrative relationship between the market and civil society? As I see it, democracy is the first main reason. As I have already stated a number of times, the ideal of sovereignty of the people as government by the people can no longer be taken literally in modern society. Bearing this in mind, I follow in the footsteps of Cohen and Arato and strike a blow for citizenship within civil society as a new approach to the ideal of sovereignty of the people. At the same time, and again in tandem with Cohen and Arato, I voice the reservation that reinterpreting the democratic ideal can only be normatively acceptable if civil society is in a position to exercise considerable influence on the state and the market. I have already discussed the possibility of influencing the state at some length. With regard to gaining influence over the market, I see the opening up of a new political-administrative channel between market and civil society as the best way forward.

The second main reason for opening a political-administrative channel between civil society and the market is the need for more prudent governance. In this study the problem of limited prudence is interpreted as the problem that, in modern-day Western societies, the idealistic component of government is far outweighed by the strategic component. This shifting balance is largely due to the increasing power of market parties in relation to the diminishing power of the state. Opening a new political-administrative channel between the market and civil society can help counteract this process. A third player is brought in, which can support the state on some occasions and the market on others. Operating on the idea that civil society is, in principle, the most rational sphere in modern-day society, it is reasonable to assume that the contribution of this third player will, in most cases, reinforce the idealistic component of government. The chance of this happening becomes even greater the more determined we are in our efforts to make present-day civil societies truly vital and rational civil societies.

I will now attempt to lend additional substance to the idea that a new political-administrative channel between civil society and the market has to be opened. Albert Hirschman's theory of feedback and recuperation mechanisms within and between organizations can be helpful in this regard. In his famous book *Exit, Voice and Loyalty* (1970) Hirschman argues that the quality of services and products always threatens to decline in economies where sheer survival is no longer at stake. Under such circumstances he regards 'slack', or diminished effort, as a natural reaction. Given this state of affairs, it is worth asking how we can prevent this loss of quality (or decline in production) from taking on unacceptable proportions. From an analytical point of view, Hirschman identifies two recuperation mechanisms present within modern society: 'exit' and 'voice'. Exit stands for breaking off the relationship between ego and alter: ego refuses to make any more use of the services or the products of alter. In the case of voice, ego does not terminate the relationship. Instead she makes her dissatisfaction known to alter, with the aim of changing alter's behaviour. According to Hirschman, exit is typical of conduct in the market, while voice is the mechanism that individuals adopt in relation to the state. (At least that is what Hirschman thought in 1970. In his later work he relaxes the relationship between the response mechanisms and the fundamental institutions, see Hirschman, 1992.)

The working of both mechanisms can be illustrated by a dissatisfied customer reaction to the quality of bread. If ego opts for exit, she will go in search of another baker. If she chooses voice, then she will complain to the baker in question. As Hirschman sees it, exit and voice do not always work well independently of one another, especially when it comes to public matters. Exit without voice always leaves alter in the dark about the nature of ego's dissatisfaction. Was the bread burnt, not big enough or too salty for ego's liking? Exit is also a costly option for ego, as she will then have to go in search of a new baker, one who may turn out to be more expensive or less conveniently located.

Meanwhile, voice without exit is often not effective. The willingness of alter to listen to ego and take appropriate action is often proportionate to the extent to which alter depends on ego. If ego cannot take a stand, for example because alter is the only baker for miles, then the latter can shrug off the complaint without any fear of consequences. In many cases voice with the threat of exit tends to be the most effective approach, where possible. The person who has the option of exit at her disposal but who first tests the waters with voice, combines the best of both worlds. By first relying on voice, ego can test alter's willingness to smarten up her act. The situation in which alter mends her ways, is by far the simplest option for ego. Because ego can brandish the powerful threat of exit at alter, there is a strong likelihood that alter will respond to her complaint. Hirschman illustrates his theory with an investigation into the quality of high schools in the United States. According to Hirschman, the research shows that the best schools are those where the parents, in addition to voice, can also ultimately take the exit option. The threat of exit keeps the school board on its toes.

Hirschman's theory does not relate directly to controlling the market. Yet it is clear that his theory is valuable in relation to this issue. The process of stimulating market parties to take some responsibility for public issues can after all be seen as an ongoing process of recuperation, which in this case is not a response to 'slack', but has to do with changes in social circumstances and social insights. Following this line of reasoning, we can infer that if civil society is to become adequately involved in the process of controlling the market, then a *political-administrative voice relationship* and a *political-administrative exit relationship* will have to be created between civil society and the market. There has to be room for effective communication between civil society organizations and market parties, and civil society's representatives have to possess the means to ensure that they are taken seriously by market parties.

How can these voice and exit relationships be given a more concrete form? In order to produce a solid picture of the voice relationship it is first necessary to come up with a more precise definition. It is possible to interpret voice as the possibility open to ego to influence alter, regardless of the reasonableness of ego's wishes. It is also possible to interpret voice as ego's opportunity to begin a serious dialogue with alter. In the latter case, voice represents the possibility of communicative action. In his work, Hirschman does not make it clear which of these two interpretations of voice he considers more adequate. In the context of this study, in which voice is seen as encouraging the prudence of market action, I therefore interpret voice as the possibility for communicative action. Given this interpretation, it is not difficult to indicate what establishing a political-administrative voice relation implies: the communication between market parties and parties from civil society has to comply with the conditions of Habermas's domination-free discourse. On the one hand this means that all parties should exhibit a willingness to enter into a mutual discussion on public issues. On the other hand, it means that all manner of technical demands must be satisfied. A very important requirement in this context is symmetry of information. In order to bring about a good voice relation between civil society and market, market parties must be prevented from exploiting their advantages in terms of information.

Even in theory it is difficult to imagine how a political-administrative exit relationship can be constructed for civil society. The clashes with the ideal of a competitive order are bound to be plentiful and severe. Nevertheless, Christopher Stone (1975) has taken up the challenge of developing ideas which are relevant in this context. According to Stone it would be a good idea to appoint a director of public affairs within businesses. The task of the director of public affairs would be to keep a finger on the pulse regarding the public consequences of the business's activities from an independent management position. Stone envisages the appointment of a director of public affairs as a procedure within which the business preselects a number of candidates and a committee acting in the interests of the state makes the final choice. Stone also gives businesses the right to fire their director of public affairs.

If it is to be useful within my context, Stone's proposal needs some adapting. To start with it seems to me preferable to entrust the director's appointment to the hands of a committee from civil society (something to which Stone explicitly and strenuously objects). In addition it is also worth asking whether a director of public affairs should be given a task within each individual corporation or within the structures at meso level (see previous section). After all, taking responsibility for public issues often involves flouting the boundaries of market discipline.

However, even with such adaptations, I still have my doubts as to whether Stone's proposal can fit the bill. Appointing a director of public affairs within businesses as a means of constructing an exit relationship for civil society is dogged by a number of significant disadvantages. In practice it is difficult for externally appointed officials to hold their own in such a position. There is a considerable chance that such an official will either come to identify with the company's interests ('capture') or be pushed into a position of isolation, as Stone himself indicates (see Stone, 1975: 153 et seq.). In my opinion, the most important argument against the idea of public directors, however, is that it does not sit easily with the ideal of a self-limiting utopia. Appointing directors of this kind demands too much involvement from civil society in concrete decision-making within the market. From this viewpoint, an indirect approach would seem more reasonable. Taking the ideal of the self-limiting utopia as a starting point, the primary aim is not to allow the citizens to participate in governance within all fundamental institutions. The main concern is that the actions taken within those fundamental institutions should be adequate from a public point of view. The influence of civil society on the market is not so much a goal in itself but a means by which to achieve this aim.

The assessment of businesses by a body on which representatives of civil society serve is a set-up which is probably more compatible with the ideal of the self-limiting utopia. By the idea of an assessment I mean that businesses should be obliged in one way or another to obtain 'proof of good conduct' or a 'licence to operate'. The process of assessment could take into account both organizational aspects (openness, communicative attitude and the like) and actual action taken. To my mind, this device for creating an exit relationship is preferable to the notion of appointing public representatives within private companies. In my opinion the most important advantage is that it makes do with a general assessment of a corporation. There is no need to monitor and check concrete decision-making processes in detail. The idea that market parties ought to be free on the market therefore remains largely intact.

I am the first to admit that the chances of this proposal being implemented anytime soon are very small. Nevertheless, I want to stress that such an approach is not completely utopian either. There are organizations within present-day civil society that are actually working towards the assessment of corporations. In the Netherlands, for example, we see that the *Kritische Konsumentenbond* (critical consumer association) has decided to assess companies on their performance in terms of social responsibility, a strategy that originated in the United States. Meanwhile, some corporations have established representative bodies from civil society for the purpose of evaluating the company's conduct. One noteworthy example is the Nestlé Infant Formula Audit Commission (NIFAC), established by Nestlé in 1982 (See: Gerber, 1990). NIFAC was given the explicit task of evaluating complaints and allegations about Nestlé's marketing practices with regard to infant formula. Of course, one should consider the emergence of NIFAC in the context of the fierce dispute on infant formula from the end of the 1960s onwards. Besides, organizations such as NIFAC are not exactly what I had in mind. But the point I want to make is that the idea of constructing institutions from civil society to evaluate the conduct of business is not totally outrageous. In practice, there are many ways in which the idea of a free and competitive market can be organized.

5.5 LOOKING BACK

In this study two mental models of social organization have been mapped out: the indirect responsibility model and the direct responsibility model. The indirect responsibility model was the dominant mental model in the 20th century. Since the 1970s, however, this mental model has come in for an increasing amount of criticism. This study has shown that criticism to be well-founded, albeit in an adapted form. The next step is to focus on the direct responsibility model, which has been worked out in far less detail, and look into the question of what form this alternative model might take.

In this chapter I have attempted to sketch the contours of the direct responsibility model, outside the confines of the three disciplines previously addressed. In doing so I have focused on the links between the fundamental institutions. In this last look back I wish to present an overall picture of what I think the direct responsibility model should look like. In doing so I will also incorporate aspects of the direct responsibility model which have emerged in previous chapters. I will also indicate points at which I feel the direct responsibility model needs to be worked out in greater detail. Below I have organized the main characteristics of the direct responsibility model in parallel with the overview of the indirect responsibility model from Section 3.2 (and 3.7).

* *The market is considered to be a sphere in which actors bear a certain responsibility for public issues.*
   This notion forms the core of the direct responsibility model. It is based on three important considerations. The first is that the state cannot cope with the burden of exclusive responsibility for public issues in a society where the collective dimension of action is large.
   Secondly the idea of public responsibility for market parties provides a solution to the paradox of planning and non-planning within the liberal-democratic tradition. At the heart of this problem lies the contradiction that on the one hand the market has to be more emphatically moulded and designed, while on the other hand the state is only permitted to exert control by means of limiting conditions. The idea that market parties carry a certain responsibility for public issues offers a solution to this problem. Planning then becomes an obligation for the market parties themselves.
   The third main reason for the transfer of a measure of responsibility for public issues to market parties has to do with the ambivalent view of the state, which so typifies the liberal-democratic tradition. On the one hand the state is supposed to be powerful, while on the other hand these powers need to be reined in. The idea that market parties ought to bear a certain responsibility for public issues makes it possible, even in a time characterized by an extensive collective dimension, to strike a balance between these rival tendencies. If market parties take on a certain responsibility of their own, limiting the state's radius of action does not necessarily clash with the need to exert adequate control over the market.
   There is still quite a bit of work to be done on the direct responsibility model in this regard. Thus far, the boundaries of the public responsibility of market parties have not been clearly enough defined. When this task is addressed, a distinction needs to be made between the individual responsibility of market parties and the collective responsibility of organized groups of market parties (branches of industry etc.). Moreover, the model needs to be differentiated with regard to the severity of the discipline imposed by the market.

* *The state is freed from the almost exclusive responsibility for public issues.*
   From the first characteristic of the direct responsibility model it follows that this mental model breaks through the strict differentiation of tasks which is such an intrinsic feature of the indirect responsibility model. In the direct responsibility model, the state no longer bears exclusive responsibility for public issues. The market and civil society also have a role to play. Of course, this is not to say that the relevance of the state for modern society diminishes. The state remains an essential fundamental institution.
   I admit that the direct responsibility model needs to be worked out in greater depth in this respect too. The question of where the boundaries of state responsibility lie should be settled more emphatically.

* *The state controls the market by imposing limiting conditions.*

Advocates of co-management distance themselves from this characteristic of the indirect responsibility model. In their eyes, the state has to learn to take a different approach to controlling the market. My interpretation of the direct responsibility model retains the idea that the state may only control the market by means of limiting conditions. As I see it, the idea of control by limiting conditions is too valuable from a normative perspective to be abandoned.

* *The state ought to be a fettered giant.*
In the field of political philosophy, every now and then figures emerge who want to free the state from its restrictions. Due to the scale of public issues, these philosophers feel that the vigour with which the state can act is in need of a substantial boost. However, the notion of an unfettered state clashes with the essence of liberal-democratic thinking. This notion therefore has no place in my vision of the direct responsibility model. It might even be said that I reinforce the idea of the fettered giant by imposing boundaries on the nature of the topics in which the state can intervene.

* *The distinction between public* issues *and private* issues *is separated from the distinction between the public* domain *and the private* domain.
The distinction between the public and the private is fundamental to liberal-democratic thinking. In this regard we can make a distinction between public *domain* versus private *domain* and public *issues* versus private *issues*. A characteristic feature of the indirect responsibility model is that these two dichotomies overlap. This overlap is one of the main reasons why the indirect responsibility model has become unfeasible as a mental model of social organization under present-day circumstances. Given the collective dimension of present-day activities there are all kinds of public issues which have their origins in the private domain. If we hold to the idea that the private domain coincides with the sphere of private issues then it becomes impossible to control these public issues adequately. The idea that these two dichotomies coincide therefore cannot be maintained.

My vision of the direct responsibility model does not interfere with the fundamental character of the distinction between private and public within liberal-democratic thinking. However, it does involve the separation of the two dichotomies, thereby creating enough scope for the adequate control of the market under present-day conditions. As I see it, the market remains part of the private domain, which means that the state is only permitted to take specific forms of action with regard to the market. However, this does not imply that market parties are only allowed to or obliged to concern themselves with private issues. In my version of the direct responsibility model the market is a sphere in which actors bear responsibility for public issues.

* *There are few bridging or linking institutions between the fundamental institutions.*
The criticism of the indirect responsibility model is often levelled at the modest amount of interaction it permits between the fundamental institutions. The advocates of co-management, for example, argue in favour of a certain measure of fusion between civil society and the state, and the market and the state. The

former is to come about by all manner of consultation by the citizens, the latter by means of corporatist structures.

Like the indirect responsibility model, my interpretation of the direct responsibility model holds to the idea that distance between the fundamental institutions is desirable. In this way all fundamental institutions are best able to realize their own logic and their own responsibilities. The term 'self limiting utopia' gives normative expression to this ideal.

*   *Democracy as citizenship in an influential civil society*

Within the indirect responsibility model, democracy refers to the sovereignty of the people. In the direct responsibility model this ideal is tempered somewhat. Democracy primarily means that the citizens are able to take an active part in the processes of public opinion-forming in civil society. As I have made clear, this position does not imply that achievements like the right to vote or existing forms of consultation should be sacrificed. Instead my position here should primarily be seen as a counterweight to limitless democratization. Secondly, it implies that the influence of civil society on the market and the state should be considerable. In my vision of the direct responsibility model, the influence of civil society on the state should be increased by the introduction of functional representation. The influence of civil society on the market should be enlarged by creating a new voice relationship and a new exit relationship between the two.

Given that civil society, as I see it, is the most rational sphere, increasing the influence of civil society on the other fundamental institutions will not only lead to greater democracy but also to greater prudence in administrative action.

In Chapter 3 the problems associated with the indirect responsibility model were systematically reduced to four points: overload of the state, lack of prudence, lack of democracy and rival tendencies. If the direct responsibility model is to be considered an improvement, then it should at least score better on a number of these points. This would indeed appear to be the case. In any event, the problem of overload of the state appears to be dealt with effectively by the direct responsibility model. The burden on the state is lightened by the distribution of responsibility for public issues over other fundamental institutions.

The problem of lack of democracy is also taken care of under the direct responsibility model. On the one hand this is achieved by redefining the term democracy for modern-day society; on the other hand the direct responsibility model attempts to enlarge the influence of civil society on the other fundamental institutions.

The problem of limited prudence is primarily dealt with in the direct responsibility model by means of the ideal of the self-limiting utopia. The ideal of the self-limiting utopia is the goal of allowing each of the fundamental institutions to work according to its own logic. This objective serves to obstruct any attempts at fusion, which generally tend to have a negative effect on the prudence of a fundamental institution. But the direct responsibility model does not pin all its hopes on differentiation in this regard. It also tries to increase the influence of (the rationality of) civil society on the other fundamental institutions.

Where the rival tendencies of liberal-democratic thinking are concerned, it is reasonable to expect that, under present-day circumstances, the direct responsibility model is more likely to succeed in reconciling them than the indirect responsibility model. The tension between planning and non-planning, and between a powerful state and a state with limited powers, are reduced under the direct responsibility model by the idea of public responsibility in the market and the idea that the private *domain* is not necessarily the domain of private *issues*.

Of course, we cannot simply conclude on the basis of this analysis alone that the direct responsibility model is superior to the indirect responsibility model. So far we have only confronted the direct responsibility model with the weaknesses of the indirect responsibility model. For a full assessment we have to carry out a comprehensive comparison of the two models as systems, which means among other things that we need to confront the direct responsibility model with its own weaknesses as well. The further analysis of the direct responsibility model, however, lies beyond the scope of this study. Nevertheless I would like to devote the final moments of this study to two weaknesses.

The first weakness concerns the motivation of actors on the market. A critic of the direct responsibility model could argue that the model overstates the degree to which people are *willing* to act responsibly in the market. At first glance, this criticism appears valid, since the direct responsibility model does indeed make a considerable appeal to this willingness. However, I doubt whether this appeal can be regarded as excessive. If individuals were only motivated by (self-regarding) preferences, then direct responsibility would indeed be too much to ask, if not downright naive. But people are also motivated by other incommensurable types of motivation. The hold exerted by these other factors, such as duty and loyalty, is actually quite strong (see for example: Etzioni, 1988; Mansbridge, 1990). It is also important to remember that for the last century acting out of self-interest within the confines of the law has been considered normal and even desirable conduct in the market. On the one hand, this implies that people can hardly be expected to act otherwise but on the other hand, it validates the proposition that people generally tend to obey the common demands of morality.

Furthermore, any mental model of social organization is bound to have its own weaknesses. When evaluating a mental model, it would be unreasonable to demand that it have no pitfalls whatsoever. However, it is reasonable to demand that the model is explicit about its weaknesses and attempts to compensate for them to a certain degree. In this chapter I have tried to deal with the stated weakness by searching for institutional devices to reinforce the position of the actor who is willing to shoulder his fair share of public responsibility. I have restricted this quest to the level of the links between the market, the state and civil society and I think that several of the suggestions I have made reinforce the position of the well-intended actor. With regard to the link between the state and the market, for example, I have looked for ways to retain competitiveness in an economic order that allows for some level of coordination between actors at the meso level. I believe that this combination of institutional devices helps the performance of the well-intended actor. With regard to the link between market and civil society, I have stressed the need for new exit and voice relations. Again, I think the well-intended market actor will benefit from this approach.

The second weakness of the direct responsibility model dovetails with the previous one. It concerns the *ability* to take responsibility for public issues in the market. Formulated as an argument, it reads as follows: the idea that market parties should take responsibility for public issues is incompatible with the idea of a competitive order. Even if market actors were willing to take responsibility, it would be impossible for them to do so because the logic of the market would not allow it. In the free market, responsibility and competition are bound to clash. I will look into this argument by distinguishing four possible ways of interpreting it.

The first interpretation is that taking responsibility for public issues puts a market party at a competitive disadvantage. To put it bluntly, the responsible businessman will be the first to go bankrupt. As I see it this is not the most interesting interpretation of the argument. Individual market parties in present-day markets as they actually exist always have a certain discretion and therefore freedom to take responsibility for public issues. Of course the freedom in a well-functioning market is limited. However, this is not to say that taking responsibility in the market is therefore impossible. Rather the objection implies that market parties can only take a substantial responsibility for public issues if they are able to coordinate their actions in one way or another. The greatest potential weakness of the direct responsibility model arises with regard to precisely this point: the question of whether the idea of a competitive order is compatible with the coordination of action on the market. This, the most interesting aspect of the criticism of the direct responsibility model, remains out of sight in the first interpretation.

A second interpretation of the argument is that structures in the market are incompatible with the desire of businesses for freedom on the market. This interpretation has no historical basis, however. It is far from true to claim that businessmen have always been the great defenders of competition (Smith, 1776b: 117). Nor is such a claim well-founded from a systematic point of view. After all, in a well-functioning market, competition is there to serve the interests of the consumers. Business on the whole has far less to gain from competition, since it forces them to provide the best possible quality at the lowest possible price. In other words, in a market where the differences in power between producers are small, most producers do not benefit from tough competition. In such a situation only the best (by definition a minority) win, while the rest go under or are forced to operate beyond their capacity. The second interpretation of this argument can therefore not be called terribly convincing.

A third interpretation of the argument might be that coordination between market parties collides head-on with the idea of a competitive order; that competition will disappear completely as a result of coordination of market action. This interpretation strikes me as too radical. Coordination of action between market parties can influence the extent of competition but this is not to say that competition will therefore be undermined completely. Even within the heavily regulated system of guilds during the Middle Ages there was still competition between guild members. Some masters were more successful than others.

This brings us automatically to the fourth interpretation: coordination of action in the market forms a threat to the market as a competitive sphere. This interpretation seems to be the most persuasive. Coordination of action on the market always implies the danger that competition will be put on a back burner, even if this is not the primary purpose of such coordination. The direct responsibility model will have to solve this problem: it will have to find a way to combine certain forms of coordination of market action with the idea of a competitive order. On a theoretical level this doesn't seem to me to be such a difficult assignment. The processes of coordination of action and the agreements made will have to be evaluated by an official body in terms of their compatibility with the idea of a competitive order. On a practical level, however, the problems are greater. Evaluating coordinated decision-making in terms of its effect on the competitiveness of the economic order might prove an onerous task.

Still, I cannot see how the formation of such a body in itself clashes with the idea of a competitive order. As things stand, many liberal democratic societies have already installed an agency to watch over the competitiveness of the market order. Similar bodies even exist at an international level. Advocates of the direct responsibility model could perhaps learn a lot from examining the actual workings of some of these bodies. The WTO might be a case in point. This organization is hampered by contradictions greatly resembling those that haunt the direct responsibility model. On the one hand the WTO forbids states to erect trade barriers against import products. On the other hand, the WTO recognizes the rights of states to draw up their own independent regulations, for example with regard to public order or the environment. This contradiction creates the need for a forum to consider whether national legislation in specific cases is in accordance with the principles of the WTO or whether it should be seen as a covert form of trade protection. Until now, the forum of the WTO has consistently declared every such instance of national legislation as being out of bounds (Brouwer, 2000). From a systematic point of view this matters little. The direct responsibility model stands in need of precisely such a forum and its advocates might therefore learn a lot about the organization of such a forum by considering this example.

As attempts to give shape to the direct responsibility model progress further, new weaknesses in this model are almost certain to appear. For now, however, I lay my analysis on this issue to rest. In any case I hope to have demonstrated in this study that thinking about the direct responsibility model is both necessary and desirable, given the dysfunctionality of the indirect responsibility model in our present-day circumstances. In addition to this, I hope I have succeeded in dispelling the myth that it is either utopian or nonsensical to expect parties in the market to bear a certain responsibility for public issues. The 'market' is far from being a timeless concept, independent of context. The idea of limited public responsibility is therefore not necessarily bound up with the idea of the market. The idea of a competitive order can be realized in various different ways. Francis Fukuyama's argument (1989: 15-20) that the fall of communism signalled 'the end of history' is therefore incorrect. The history of the liberal democracy is set to continue into the 21st century.

APPENDIX

AN OVERVIEW OF THE DIRECT AND THE INDIRECT RESPONSIBILITY MODELS

	The indirect responsibility model	*The direct responsibility model* (Aspects of the model which the author sets out to strengthen in Chapters 4 and 5 are indicated as 'weak points'.)
Degree of maturation	Ideology (in Mannheimian sense)	Utopia (in Mannheimian sense)
View of the need to control the market in dealing with public issues	There is a strong need to control the market in this regard.	There is a strong need to control the market in this regard.
View of the differentiation of market, state and civil society within modern society	The desirability of differentiation between the spheres is stressed.	The problems of strong differentiation are stressed. (*The author sees this one-sidedness as a weak point.*)
View of the differentiation within the fundamental institutions (especially the market)	The desirability and need for differentiation within the fundamental institutions is stressed.	The desirability of a certain amount of dedifferentiation in the market and civil society is stressed.
View of the fundamental institutions involved in controlling the market	Controlling the market is (almost) exclusively a responsibility of the state.	Market, state and civil society each have a responsibility.
View of the need to clearly allocate responsibilities	The need for clear demarcation of responsibilities between state, market and civil society is stressed.	The need for clear demarcation of responsibilities is not stressed. (*Weak point in the author's view.*)
View of the potential power of the state (in relation to controlling the market)	The subject is not given much consideration but the potential power of the state is deemed to be great. The state can be used to deal with any corrective task.	There are systematic limits to the power of the state. The state should therefore be looked upon as a limited-use institution.
View of civil society (in relation to controlling the market)	Civil society is (almost) irrelevant in this regard.	The model tries to find new ways to make civil society relevant in this regard. (*The author regards the model's elaboration of this point as weak.*)
View of the way in which the market ought to be controlled.	Extreme institutionalism The state ought to control the market at system level by limiting its conditions.	A mix of institutionalism and voluntarism Actors are also called upon to personally acknowledge responsibility.

View of the desirability of fusion between the fundamental institutions in relation to the administrative process of controlling the market	Fusion is not desirable. 'Command and control' instruments are deemed fitting.	It is suggested that a certain amount of fusion between state, market and civil society is desirable. The desirability of 'communicative instruments' is exclusively stressed. *(Weak point according to author.)*
View of the general relation between market and morality	The market is conceptualized as a moral sphere where actors are required to abide by the law and the rules of common decency.	The moral duties of market actors do not begin and end with their legal duties.
View of the responsibility of market actors in relation to dealing with public issues	If a market actor has any public responsibility, it is confined to their republican duty.	The public responsibilities of a market actor exceed their republican duty.
View of the relation between the public and the private as 'domains' and as 'issues'.	The two components of the distinction between the public and the private are superimposed upon each other.	The two components of the distinction between the public and the private are (implicitly) separated.
View of the motivation of actors in the market	Market actors act out of self-interest (within the confines of the law).	The economic value of acting in the public interest is stressed. The theoretical need for a new view of motivation is not articulated. *(Weak point according to author.)*
View of the compatibility of market forces (competitiveness) and effective control of the market	Competitiveness is compatible with control by limiting conditions.	The subject is not elaborated upon. *(Weak point according to author.)*
View of the value of the market	The market is valued as an instrument (i.e. in terms of its functionality).	The subject is not elaborated upon. *(Weak point according to author.)*
View of democracy	Democracy is reduced to representation within the state.	The subject is hardly elaborated upon. *(Weak point according to author.)*
View of the desirability of restraining the powers of the state	The need for restraining the powers of the state is acknowledged.	The subject is not elaborated upon. *(Weak point according to author.)*
View of rival tendencies with regard to planning	The rival tendencies are balanced by restricting planning to the macro-level.	The subject is hardly elaborated upon. *(Weak point according to author.)*

BIBLIOGRAPHY

- Aalberse, P.J.M. and H. Pesch s.j. *Opkomst, bloei en verval der gilden.* (Uitgevers Vennootschap 'Futura') Leiden, 1912.
- Aalders, A. *Handhaving van milieurecht.* (Boom) Amsterdam, 1987.
- Acheson, J.A. and B.J. McCay (eds.) *The Question of the Commons. The Culture and Ecology of Communal Resources.* (University Press of Arizona) Tucson, 1990.
- Acheson, J.H. *The Lobster Gangs of Maine.* Hannover, 1988.
- Achterberg, W. 'Leven te midden van leven / Fragment van een milieu-ethiek.' In: *Filosofie en Praktijk* IV/2 (1983) 57-76.
- Achterberg, W. and W. Zweers (eds.) *Milieucrisis en filosofie.* (Ekologische Uitgeverij) Amsterdam, 1984.
- Achterberg, W. and W. Zweers (eds.) *Milieufilosofie tussen theorie en praktijk.* (Jan van Arkel) Utrecht, 1986.
- Achterberg, W. *Samenleving, natuur en duurzaamheid. Een inleiding in de milieu-filosofie.* (Van Gorcum) Assen, 1994.
- Albert, M. *Kapitalisme contra kapitalisme.* (Contact) Antwerp, 1992.
- Albrow, M. *Bureaucracy.* (MacMillan) London, 1970.
- Alec Gee, J. 'The Neoclassical School.' In: D. Mair and A. Miller (eds.) *A Modern Guide to Economic Thought.* (Edward Elgar) Brookfield, 1991: 71-108.
- Andersen, M.S. *Governance by Green Taxes. Making Pollution Prevention Pay.* (Manchester University Press) Manchester, 1994.
- Anderson, L.G. *The Economics of Fisheries Management.* () Baltimore, 1977.
- Andeweg, R. 'Overheid of overhead? De bestuurbaarheid van het overheidsapparaat.' In: M. Bovens and W. Witteveen (eds.) *Het schip van staat. Beschouwingen over recht, staat en sturing.* (Tjeenk Willink) Zwolle, 1985: 207-214.
- Ankersmit, F.R. 'Tocqueville en de ambivalentie van de democratie.' In: T. Kuipers (ed.) *Filosofen in actie.* (Eburon) Groningen, 1992: 119-132.
- Ankersmit, F.R. 'Democratie als conflictbeheersing.' In: *Krisis* ./52 (1993) 5-19.
- Ankersmit, F.R. 'Politieke partijen in het tijdperk van de onbedoelde gevolgen.' In: *Jaarboek documentatiecentrum Nederlandse politieke partijen 1994.* Groningen (1995) 207-237.
- Ankersmit, F.R. *Macht door representatie. Exploraties III: politieke filosofie.* (Kok Agora) Kampen, 1997.
- Anshen, M. *Managing the Socially Responsible Corporation.* (MacMillan) New York, 1975.
- Arrow, K.J. 'Social Responsibility and Economic Efficiency.' In: *Public Policy* XXI. (1973) 303-317.
- Asperen, T. 'Met de beste bedoelingen.. . Over de ideologie van de verzorgingsstaat.' In: *Filosofie en Praktijk* II/3 (1981) 167-181.
- Asperen, T. 'Twee democratische tradities.' In: *Filosofie en Praktijk* V/1 (1984) 30-39.
- Axelrod, R. *The Evolution of Cooperation.* (Basic Books) New York, 1984.
- Backes, Ch. and P.J.J. Buuren 'Loopt de Grote Peel leeg?' In: P.J.J. Buuren and G. Betlem *Milieurecht in stelling. Utrechtse opstellen over actuele thema's in het milieurecht.* (W.E.J. Tjeenk-Willink) Zwolle (1990) 151-157.
- Baden, J. and R.L. Stroup (eds.) *Bureaucracy vs. Environment. The Environmental Costs of Bureaucratic Governance.* () Ann Arbor, 1981.
- Bader, V.M. 'Schmerzlose Entkopplung von System und Lebenswelt?' In: *Kennis en Methode* VII/. (1983) 329-355.
- Bailyn, B. *The Ideological Origins of the American Revolution.* (Belknap Press of Harvard University Press) Cambridge Mass., 1967/1982.
- Banning, W. *Om mens en menselijkheid in maatschappij en politiek.* (Meulenhoff) Amsterdam, 1960.
- Barber, B. *Strong Democracy: Participatory Politics for a New Age.* (University of California Press) Berkeley, 1984.
- Barbier, B., M. Acreman and D. Knowler *Economic Valuation of Wetlands. A Guide for Policymakers and Planners.* (Ramsar Convention Bureau) New York, 1997.

- Bardach, E. and R.A. Kagan *Going by the Book. The Problem of Regulatory Unreasonableness.* (Temple University Press) Philadelphia, 1982.
- Barnet, R.J. *The Lean Years: Politics in the Age of Scarcity.* (Simon and Schuster) New York, 1980.
- Barrett, G. and T. Okudaira 'The Limits of Fishery Cooperatives? Community Development and Rural Depopulation in Hokkaido, Japan.' In: *Economic and Industrial Democracy* XVI/. (1995) 201-232.
- Barry, B. *Political Argument. A Reissue with a New Introduction.* (University of California Press) Berkeley, 1965/1990.
- Baumol, W. 'Business Responsibility and Economic Behavior.' In: M. Anshen *Managing the Socially Responsible Corporation.* (MacMillan) New York (1975) 59-74.
- Baumol, W.J. and S.J. Batey Blackman *Perfect Markets and Easy Virtue: Business Ethics and the Invisible Hand.* (Basil Blackwell) Oxford, 1991.
- Beck, U. *Risikogesellschaft.* (Suhrkamp) Frankfurt am Main, 1986.
- Beck, U. *Gegengifte - Die organisierte Unverantwortlichkeit.* (Suhrkamp) Frankfurt am Main, 1988.
- Beck, U. *Die Erfindung des Politischen. Zu einer Theorie reflexiver Modernisierung.* (Suhrkamp) Frankfurt am Main, 1993.
- Becker, M. and K. Klop et al. (eds.) *Economie en ethiek in dialoog.* (Van Gorcum) Assen, 2001.
- Beiner, R. *Theorizing Citizenship.* (State University of New York Press) Albany, 1995.
- Bell, D. *The Coming of Post-Industrial Society: A venture in Social Forecasting.* (Basic Books) New York, 1973.
- Bentham, J. (and W. Harrison) *A Fragment on Government and an Introduction to the Principles of Morals and Legislation.* (Basil Blackwell) Oxford, 1776/1789/1823/1948/1967.
- Bentham, J. (and M. Quinn (ed.)) *The Collected Works of Jeremy Bentham. Writings on the Poor Laws. Volume 1.* (Clarendon Press) Oxford, unpublished/ 1797/ 2001.
- Bentham, J. 'Defence of Usury.' In: J. Bentham (and W. Stark (ed.)) *Jeremy Bentham's Economic Writings. Critical Edition Based on his Printed Works and Unprinted Manuscrips. Volume 1.* (The Royal Economic Society & George Allen & Unwin) London, 1787/ 1952.
- Berkes, F. (ed.) *Common Property Resources. Ecology and Community Based-Sustainable Development.* () Bristol, 1989.
- Berkes, F., F. Foeny, B.J. McCay and J.M. Acheson 'Reassessing the Commons.' *typoscript.*
- Berle, A.A. and G.C. Means *The Modern Corporation and Private Property.* (MacMillan) New York, 1932.
- Berns, E. 'Over de Grens van de Economie.' In: *Tijdschrift voor Filosofie ./*1 (1992) 1-16.
- Bernstein, R.J. *Beyond Objectivism and Relativism: Science, Hermeneutics, and Praxis.* Oxford, 1983.
- Bernstein, R.J. 'Dewey, Democracy: the Task ahead of Us.' In: J. Rajchman and C. West *Post-Analytic Philosophy.* (Columbia University Press) New York, 1985.
- Beus, J.W. de, and J.J.A. van Doorn (eds.) *De Interventiestaat. Tradities, ervaringen, reacties.* (Boom) Meppel, 1984.
- Beus, J.W. de, and P. Lehning (eds.) *Beleid voor een Vrije Samenleving. Beleid en Maatschappij Jaarboek 1989* (Boom) Meppel, 1989.
- Beus, J.W. de, *Markt, Democratie en Vrijheid. Een Politiek-Economische Studie.* (Tjeenk Willink) Zwolle, 1989.
- Beus, J.W. de, 'Het Optimale, het Fundamentele en het Procedurele. Een Notitie over de, Overheidstaken in de Naoorlogse Welvaartstheorie.' In: *Beleid en Maatschappij* XV/. (1988) 168-178.
- Beus, J.W. de, 'Het nieuwe rationele-keuze institutionalisme.' In: *Beleid en Maatschappij* XXI/. (September/October 1994) 246-260.
- Biesboer, F. *Greep op groei. Het thema van de jaren '90.* (Aktie Strohalm) Utrecht, 1993.
- Bijker, W.E. and J. Law (eds.) *Shaping Technology/Building Society. Studies in Sociotechnical Change.* (MIT Press) Cambridge Mass., 1992.
- Bijker, W.E. 'Politisering van de technologische cultuur.' In: *Kennis en Methode* XX/3 (1996) 294-307.
- Blansch, K. le 'De zin van hedendaags beleid. Bedrijfsmilieuzorg als voorbeeld.' In: *Beleid en Maatschappij* XXI/2 (1995) 22-31.
- Blaug, M. *Economic Theory in Retrospect.* Fifth edition. (Cambridge University Press) Cambridge, 1992/1996.
- Boatright, J.R. *Ethics and the Conduct of Business.* Third edition. (Prentice Hall) Upper Saddle River, 1993/2000.
- Boatright, J.R. 'Does Business Ethics Rest on a Mistake?' In: *Business Ethics Quarterly* IX/4 (1999) 583-592.

- Bobbio, N. *The Future of Democracy.* () Cambridge, 1984/1987.
- Borg, M.B. ter, 'Op Zoek naar de Ideologie van de Verzorgingsstaat.' In: *Filosofie en Praktijk* VI/2 (1985) 66-79.
- Bormans, P. and T. Hamers *Milieukunde.* (Spectrum) Utrecht, 1985.
- Bovens, M.A.P. and W.J. Witteveen (eds.) *Het schip van Staat. Beschouwingen over recht, staat en sturing.* (Tjeenk Willink) Zwolle, 1985.
- Bovens, M.A.P. *Verantwoordelijkheid en Organisatie. Beschouwingen over Aansprakelijkheid, Institutioneel Burgerschap en Ambtelijke Ongehoorzaamheid.* (Tjeenk Willink) Zwolle, 1990.
- Bovens, M.A.P., W. Derksen et al. *De Verplaatsing van de Politiek. Een Agenda voor Democratische Vernieuwing.* (Wiardi Beckman Stichting) Amsterdam, 1995.
- Bovens, M.A.P. 'De centrumloze democratie en het primaat van de politiek.' In: *Openbaar Bestuur* ./1 (1996) 2-7.
- Bovens, M.A.P. 'Politiek-ambtelijke verhoudingen.' In: U. Rosenthal, A.B. Ringeling and M.A.P. Bovens *Openbaar bestuur: beleid, organisatie en politiek.* Fifth fully revised edition. (Samsom H.D. Tjeenk Willink) Alphen aan den Rijn, 1996.
- Bovens, M.A.P. 'De maatschappelijke omgeving van het openbaar bestuur.' In: U. Rosenthal, A.B. Ringeling and M.A.P. Bovens *Openbaar bestuur: beleid, organisatie en politiek.* Fifth fully revised edition. (Samsom H.D. Tjeenk Willink) Alphen aan den Rijn, 1996.
- Bowie, N. 'Morality, Money, and Motor Cars.' In: W.M. Hoffman, R. Frederick and E.S. Petry, jr. *Business, Ethics, and the Environment. The Public Policy Debate.* (Quorum Books) New York (1989) 89-97.
- Bowie, N. *Business Ethics. A Kantian Perspective.* (Blackwell Publishers) Malden, 1999.
- Bowles, S. and H. Gintis *Democracy and Capitalism: Property, Community and the Contradictions of Modern Social Thought.* (Basic Books) New York, 1986.
- Braam, A. van (ed.) *Sociologie van het staatsbestuur. Two Volumes.* (Universitaire Pers Rotterdam/Standaard) Rotterdam, 1969.
- Braams, W.Th., F.W. Grosheide and E.A. Messer 'Handhaving door algemeen-belangbehartigers via het milieuaansprakelijkheidsrecht.' In: P.J.J. Buuren and G. Betlem *Milieurecht in stelling. Utrechtse opstellen over actuele thema's in het milieurecht.* (W.E.J. Tjeenk-Willink) Zwolle (1990) 1-23.
- Braithwaite, J. 'Enforced Self-Regulation: a New Strategy for Corporate Crime Control.' In: *Michigan Law Review* LIII/. (June 1982) 1466-1507.
- Brand, A.F. and H. van Luijk (eds.) *Bedrijfsethiek in Nederland.* () The Hague, 1989.
- Breiner, P. 'The Political Logic of Economics and the Economic Logic of Modernity in Max Weber.' In: *Political Theory* XXIII/1 (1995) 25-47.
- Breitenbach, H., T. Burden and D. Coates 'Socialism, Planning, and the Market.' In: G. Thompson, J. Frances et al. (eds.) *Markets, Hierarchies and Networks.* (Sage) London, 1991: 48-53.
- Bressers, J.Th.A. and P.J. Klok 'Ontwikkelingen in het Nederlandse milieubeleid.: doelrationaliteit of cultuurverschuiving?' In: *Beleidswetenschap* X/. (1996) 445-460.
- Brom, F.W.A. *Onherstelbaar verbeterd. Biotechnologie bij dieren als moreel probleem.* (Van Gorcum) Assen, 1997.
- Bromley, D.W. *Environment and Economy, Property Rights and Public Policy.* (Basil Blackwell) Cambridge, 1991.
- Brouwer, A.-M. de, '*The World Trade Organization and Ethical Concerns Regarding Animals and Nature: Is There Room for National Policies? A Legal Analysis.* (Globus) Tilburg, 2000.
- Brown, L.R. *Building a Sustainable Society.* () New York, 1981.
- Bruijn, J.A. and J.A.M. Hufen 'Instrumenten van overheidsbeleid.' In: *Beleidswetenschap* VI/1 (1992) 69-93.
- Brussaard, W., Th. Drupsteen et al. (eds.) *Milieurecht.* (Tjeenk Willink) Zwolle, 1991.
- Buchanan, J.M. 'Political Equality and Private Property: the distributional paradox.' In: Dworkin, R., G. Bermant and P.G. Brown (eds.) *Markets and Morals.* (Hemisphere Publishing Company / John Wiley and Sons) London (1977) 69-84.
- Buchanan, A. *Ethics, Efficiency and the Market.* (Clarendon Press) Oxford, 1985.
- Buckley, W., T.R. Burns and D. Meeker 'Structural Resolutions of Collective Action Problems.' In: T.R. Burns, et al. (eds.) *Man, Decision and Society.* () New York (1985) 79-111.
- Burg, W. van der, 'Het milieu in de politieke filosofie. Naar een andere staatsopvatting.' In: W. Achterberg (ed.) *Natuur: uitbuiting of respect. Natuurwaarden in discussie.* (Agora) Kampen, 1989.

- Burg, W. van der, *Het democratisch perspectief. Een verkenning van de normatieve grondslagen van de democratie.* (Gouda Quint) Arnhem, 1991.
- Buurma, H., J.B.M. Edelman-Bos and J.J. Swanink (eds.) *Management bij de overheid. Het belang van een nieuw elan.* (Nederlands Studie Centrum) Vlaardingen, 1986.
- Child, J. 'Organizational Structure, Environment and Performance.' In: *Sociology* VI/.1 (1972) 1-22.
- Coase, R.H. 'The Nature of the Firm.' In: *Economica* IV/. (1937) 386-405.
- Coase, R.H. 'The Problem of Social Costs.' In: *The Journal of Law and Economics* III/. (Oktober 1960) 1-41.
- Coase, R. 'The Firm, the Market and the Law.' In: R. Coase (ed.) *The Firm, the Market and the Law.* (University of Chicago Press) Chicago, 1988: 1-7.
- Cobden, R. *The Political Writings (In Two Volumes).* (William Ridgway) London, 1867.
- Coe, R.D. and C.K. Wilber *Capitalism and Democracy. Schumpeter Revisited.* (University of Notre Dame Press) Notre Dame, 1985.
- Cohen, J. 'An Epistemic Conception of Democracy.' In: *Ethics* ./97 (Oktober 1986) 26-38.
- Cohen, J., J. Rogers and E.O. Wright (eds.) *Associations and Democracy. The Real Utopia's Project. Volume I.* (Verso) London, 1995a.
- Cohen, J. 'Secondary Associations and Democratic Governance.' In: J. Cohen, J. Rogers and E. Wright (eds.) *Associations and Democracy. The Real Utopia's Project. Volume I.* (Verso) London, 1995b.
- Cohen, J.L. and A. Arato *Civil Society and Political Theory.* (MIT Press) Cambridge Mass., 1992.
- Coleman, J. and J. Ferejohn 'Democracy and Social Choice.' In: *Ethics* XCVII/. (October 1986) 6-23.
- Coleman, J.S. *The Asymmetric Society.* (Syracuse University Press) New York, 1982.
- Colletti, L. *From Rousseau to Lenin: Studies in Ideology and Society.* (Monthly Review Press) London, 1972.
- Commissie Bedrijfsinterne Milieuzorg *Milieuzorg in samenspel.* () The Hague, 1988.
- Commissie Lange Termijn Milieubeleid *Het milieu: denkbeelden voor de 21ste eeuw.* (Kerckebosch) Zeist, 1990.
- Commons, J.R. *Legal Foundations of Capitalism.* (MacMillan) New York, 1924.
- Couwenberg, S.W. *Het particuliere stelsel. De behartiging van publieke belangen door particuliere lichamen.* (Samsom) Alphen aan den Rijn, 1953.
- Cramer, J. and W. Zegveld 'Schoon produceren: wie kan er wat aan doen?' In: Commissie Lange Termijn Milieubeleid *Het milieu: denkbeelden voor de 21ste eeuw.* (Kerckebosch) Zeist (1990) 391-410.
- Crince le Roi, R. *De vierde macht: de ambtelijke bureaucratie als machtsfactor in de staat.* () Baarn, 1971.
- Cunningham, S., M.R. Dunn and D. Whitmarsh *Fisheries Economics. An Introduction.* London, 1993.
- Daalder, H. 'Sturing, het Primaat van de Politiek en de Bureaucratische Cultuur in Nederland.' In: M. Bovens and W. Witteveen (eds.) *Het Schip van Staat. Beschouwingen over Recht, Staat en Sturing.* (Tjeenk Willink) Zwolle, 1985: 197-207.
- Daele, W. van den, 'Het Behoud van de Schepping als Overheidstaak?' In: J. Keulartz and M. Korthals (eds.) *Museum Aarde. Natuur: Criterium of Constructie?* (Boom) Amsterdam (1997) 21-36.
- Dahl, R.A. *After the Revolution?* (Yale University Press) New Haven, 1970.
- Dahl, R.A. *A Preface to Economic Democracy.* (University of California Press) Berkeley, 1985.
- Dahl, R.A. *Democracy, Liberty and Equality.* (Norwegian University Press) Oslo, 1986.
- Dahl, R.A. *Democracy and its Critics.* (Yale University Press) New Haven, 1989.
- Dahrendorf, R. *Soziale Klassen und Klassenkonflikt in der industriellen Gesellschaft.* Stuttgart, 1957.
- Dahrendorf, R. *Plan und Markt. Zwei Typen der Rationalität.* (Mohr, Paul Siebeck) Tübingen, 1966.
- Dahrendorf, R. *The Modern Social Conflict. An Essay on the Politics of Liberty.* (Weidenfeld and Nicholson) New York, 1988.
- Daly, H.E. (ed.) *Economics, Ecology, Ethics. Essays toward a Steady State Economy.* (Freeman and Company) San Francisco, 1980.
- Daly, H.E. and J.B. Cobb jr. *For the Common Good. Redirecting the Economy toward Community, the Environment and a Sustainable Future.* (Beacon Press Books) Boston, 1989.
- Daly, H.E. *Steady-State Economics.* Second Edition with New Essays. (Earthscan) London, 1992.
- Davidse, W.P. 'Negatieve bedrijfsresultaten in de kottervisserij en toch een nieuwbouwgolf. Hoe kan dat?' In: *Visserij* XXXVIII/1 (1985) 21-26.
- Dekker, P. (ed.) *Civil Society. Verkenningen van een perspectief op vrijwilligerswerk. Civil society en vrijwilligerswerk I.* (Sociaal en Cultureel Planbureau) Rijswijk, 1994.

- Deléage, J.P. 'De ecologische grens: groen kapitalisme of nieuwe emancipatiehorizon?' In: *Oikos* I/1 (Autumn 1996) 11-21.
- Denhardt, R.D. *Theories of Public Organizations.* (Brooks/Cole Publishing Company) Monterey, 1984.
- Dewey, J. (and J.A. Boydston (ed.)) *Ethics. The Middle Works, Volume 5: 1905a* (Southern Illinois University Press) Carbondale, 1980a.
- Dewey, J. 'Civil Society and the Political State.' In: J. Dewey (and J.A. Boydston (ed.)) *Ethics. The Middle Works. Volume V: 1905b.* (Southern Illinois University Press) Carbondale, 1980b: 404-434.
- Dewey, J. 'The Ethics of Economic Life.' In: J. Dewey (and J.A. Boydston (ed.)) *Ethics. The Middle Works. Volume V: 1905c.* (Southern Illinois University Press) Carbondale, 1980c: 435-459.
- Dewey, J. 'Some Principles in the Economic Order.' In: J. Dewey (and J.A. Boydston (ed.)) *Ethics. The Middle Works. Volume V: 1905d.* (Southern Illinois University Press) Carbondale, 1980d: 460-467.
- Dewey, J. (and J.A. Boydston (ed.)) *Democracy and Education. The Middle Works, Volume 9: 1916e.* (Southern Illinois University Press) Carbondale, 1980e.
- Dewey, J. (and J.A. Boydston (ed.)) *Essays, Reviews, Miscellany and 'The Public and its Problems'. The Later Works, Volume 2: 1925-1927f.* (Southern Illinois University Press) Carbondale, 1984f.
- Dewey, J. 'The Public and its Problems.' In: Dewey, J. (and J.A. Boydston (ed.)) *Essays, Reviews, Miscellany and 'The Public and its Problems'. The Later Works, Volume 2: 1925-1927g.* (Southern Illinois University Press) Carbondale, (1984g) 235-372.
- Dewey, J. (and J.A. Boydston (ed.)) *Essays. 'The Sources of a Science of Education', 'Individualism, Old and New', and 'Construction and Criticism'. The Later Works, Volume 5: 1929-1930h.* (Southern Illinois University Press) Carbondale, 1984h.
- Dewey, J. (and J.A. Boydston (ed.)) *Ethics. The Later Works, Volume 7: 1932i.* (Southern Illinois University Press) Carbondale, 1985i.
- Dewey, J. 'Toward the Future.' In: Dewey, J. (and J.A. Boydston (ed.)) *Ethics. The Later Works, Volume 7: 1932j.* (Southern Illinois University Press) Carbondale (1985j) 423-437.
- Dewey, J. 'Creative Democracy. The Task before Us.' In: J. Dewey (and J.A. Boydston (ed.)) *The Later Works, Volume 14: 1939-1941k.* (Southern Illinois University Press) Carbondale (1986k) 224-230.
- Dobson, A. and P. Lucardie (ed.) *The Politics of Nature. Explorations in Green Political Theory.* (Routledge) London, 1993.
- Doel, J. van den, *Economie en demokratie in het staatsbestuur.* (Kluwer) Deventer, 1973.
- Doel, J. van den, *Demokratie en Welvaartstheorie.* Second edition. (Samsom) Alphen aan den Rijn, 1978.
- Donaldson, T. *Corporations and Morality.* (Prentice Hall) Englewood Cliffs, 1982.
- Dooyeweerd, H. *De wijsbegeerte der wetsidee.* (H.J. Paris) Amsterdam, 1935.
- Dooyeweerd, H. *In the Twilight of Western Thought.* (Presbyrterian and Reformed Publishing) Philadelphia, 1960.
- Dore, R. 'Goodwill and the Spirit of Capitalism.' In: Granovetter, M. and R. Swedberg (eds.) *The Sociology of Economic Life.* (Westview Press) Boulder (1983/1992) 159-180.
- Dragon, A.K. and M.P. O'Connor 'Property Rights, Public Choice, and Pigovianism.' In: *Journal of Post Keynesian Economics* XVI/1 (fall 1993) 127-151.
- Drucker, P.F. *Management. Tasks, Responsibilities, Practices.* (Heinemann) London, 1974.
- Drucker, P.F. 'The New Meaning of Corporate Social Responsibility.' In: *California Management Review.* XXVII/2 (Winter 1984) 53-63.
- Dryzek, J.S. *Rational Ecology. Environment and Political Economy.* (Basil Blackwell) Oxford, 1987.
- Dryzek, J. 'Strategies of Ecological Modernization.' In: W. Lafferty and J. Meadowcroft (eds.) *Democracy and the Environment.* (Edward Elgar) Cheltenham (1997) 108-123.
- Dubbink, W. 'Democratie-theorie en de milieuproblematiek.' *Filosofie en praktijk* XV/3 (Autumn 1994) 143-164.
- Dubbink, W. and M. van Vliet 'Market regulation versus co-management? Two perspectives on regulating fisheries compared.' *Marine Policy* XX/6 (1996) 499-516.
- Dubbink, W. 'Spelen met het milieu. Theorie en praktijk van de spel-theorie.' *Rotterdam School of Management. Management report series no. 285* (Erasmus Universiteit) Rotterdam, 1996.
- Dubbink, W. 'Dieren, dienders en democratie. De Gezondheids- en Welzijnswet voor Dieren en het probleem van de bureaucratie.' *Kennis en Methode* XXII/3 (1998) 312-338.
- Duijse, P. van, A. Nentjes, et al. *Verhandelbare CO2-emissierechten.* (VROM-raad) The Hague, 1998.
- Durkheim, E. (and H.P.M. Goddijn) *Over Moraliteit.* (Boom) Meppel, 1920/1977.
- Durkheim, E. *Professional Ethics and Civic Morals.* (Routledge) London, 1922/1992.

- Dworkin, R. *Taking Rights Seriously*. (Harvard University Press) Cambridge Mass., 1977.
- Dworkin, R. 'Liberalism.' In: S. Hampshire *Public and Private Morality*. (Cambridge University Press) Cambridge (1978/1979) 113-143.
- Eckersley, R. *Environmentalism and Political Theory. Toward an Ecocentric Approach*. (UCL-Press) London, 1992.
- Edelman-Bos, J. 'Structureren en Managen bij de Overheid: Behoefte aan een Nieuw Bestuurlijk Elan.' In: H. Buurma, J. Edelman-Bos and J. Swanink (eds.) *Management bij de overheid*. (Nederlands Studie Centrum) Vlaardingen, 1986: 17-28.
- Eijgelshoven, P.J., A. Nentjes and B.C.J. van Velthoven *Markten en overheid*. (Wolters-Noordhoff) Groningen, 1993.
- Eijlander, Ph., P.C. Gilhaus and J.A.F. Peters *Overheid en zelfregulering. Alibi voor vrijblijvendheid of prikkel tot actie?* (Tjeenk Willink) Zwolle, 1993.
- Ehrlich, P. (and A. Ehrlich) *Population, Resources, Environment*. (Freeman) San Francisco, 1972.
- Eliassen, K.A. and J. Kooiman (eds.) *Managing Public Organizations. Lessons from Contemporary European Experience*. (Sage) London, 1993.
- Elster, J. *The Cement of Society. A Study of Social Order*. (Cambridge University Press) Cambridge, 1989.
- Elster, J. 'The Market and the Forum: Three Varieties of Political Theory.' In: J. Elster and A. Hylland *Foundations of Social Choice Theory*. (Cambridge University Press) Cambridge (1986) 103-132.
- Engelen, E. *De mythe van de markt. Waarheid en leugen in de economie*. Amsterdam, 1995.
- Etzioni, A. *The Moral Dimension. Toward a New Economics*. (The Free Press) New York, 1988.
- Etzioni-Havely, E. *Bureaucracy and Democracy. A Political Dilemma*. Revised edition. (Routledge and Kegan Paul) London, 1983/1985.
- Evans-Pritchard, E.E. *The Nuer. A Description of the Modes of Livelihood and Political Institutions of a Nilotic People*. (Oxford University Press) Oxford, 1940/1969.
- Ferguson, A. (and F. Oz-Saltberger (ed.)) *An Essay on the History of Civil Society*. (Cambridge University Press) Cambridge, 1768/1995.
- Feyerabend, P. *Science in a Free Society*. (Verso) London, 1978.
- Field, B.C. 'The Evolution of Property Rights.' In: *Kiklos* XLII/3 (1989) 319-345.
- Finkelstein, J. and A.L. Thimm *Economists and Society. The Development of Economic Thought from Aquinas to Keynes*. (Harper and Row Publishers) New York, 1975.
- Foucault, M. *Discipline and Punish. The Birth of the Prison*. (Penguin Books) New York, 1977.
- Frankel, B. *The Post-industrial Utopians*. (Polity Press) Cambridge, 1987.
- Freeman III, A.M., A.V. Kneese and R.H. Haveman *The Economics of Environmental Policy*. (J. Wiley and Sons) New York, 1973.
- Friedman, M. 'The Social Responsibility of Business Is to Increase its Profits.' In: *The New York Times* (13 September 1970) 33.
- Frissen, P.H.A. and V. Bekkers 'Een afwerkplek voor de boomkikker.' In: *Openbaar Bestuur* ./8 (1997) 20-23.
- Frissen, P.H.A. *De versplinterde staat. Over informatisering, bureaucratie en technocratie voorbij de politiek*. (Samsom H.D. Tjeenk Willink) Alphen aan den Rijn, 1991.
- Fukuyama, F. 'Het Einde van de Geschiedenis.' In: *De Groene Amsterdammer* (December 20, 1989) 15-20.
- Fukuyama, F. *Trust. The Social Virtues and the Creation of Prosperity*. (Hamish Hamilton) London, 1995.
- Galbraith, J.K. *The New Industrial State*. (Penguin Books) Harmondsworth, 1967/1969.
- Galbraith, J.K. *A History of Economics. The Past as Present*. (Penguin Books) Harmondsworth, 1987.
- Galbraith, J.K. *The Affluent Society*. Fourth edition. (Penguin Books) Harmondsworth, 1958/1991.
- Gauthier, D. *Morals by Agreement*. (Clarendon Press) Oxford, 1986.
- Geelhoed, L. 'Markt, Ordening en Sturing: Schuivende Posities.' In: A. Nentjes (ed.) *Marktwerking versus Coördinatie. Preadviezen van de Koninklijke Vereniging voor de Staatshuishoudkunde 1996*. (Lemma) Utrecht (1996) 11-35.
- Gerber, J. 'Enforced Self-Regulation in the Infant Formula Industry: a Radical Extension of an "Impractical" Solution.' *Social Justice* XVII/1 (1990) 98-112.
- Geus, M. de 'Milieubeleid en democratie.' In: *Oefeningen in duurzaamheid: perspectieven naar 2040*. (Van Arkel) Utrecht, 1995.
- Gherity, J.A. (ed.) *Economic Thought. A Historical Anthology*. (Random House) New York, 1965.

- Giddens, A. *Max Weber over politiek en sociologie.* (Boom) Meppel, 1973/1974.
- Giebels, R. 'Chinees Leger is Economische Macht.' In: *NRC Handelsblad* (Thursday, February 26, 1998) 17.
- Giner, S. 'The Withering Away of Civil Society?' In: *Praxis International* IV/. (1985).
- Glasbergen, P. (ed.) *Milieubeleid, Theorie en praktijk.* (Vuga) The Hague, 1989a.
- Glasbergen, P. *Beleidsnetwerken rond milieuproblemen. Een beschrijving van de relevantie van het denken in termen van beleidsnetwerken voor het analyseren en oplossen van milieuproblemen.* (Vuga) The Hague, 1989b.
- Glasbergen, P. 'Seven Steps toward an Instrument Theory for Environmental Policy.' In: *Policy and Politics* XX/3 (1992) 191-200.
- Glasbergen, P. (ed.) *Managing Environmental Disputes. Network Mananagement as an Alternative.* (Kluwer Academic Publishers) Dordrecht, 1995.
- Glasbergen, P. 'Learning to Manage the Environment.' In: W. Lafferty and J. Meadowcroft (eds.) *Democracy and the Environment.* (Edward Elgar) Cheltenham (1997) 175-193.
- Glasbergen, P. 'De feilbare politieke modernisering.' *typoscript,* (Universiteit Utrecht) Utrecht, 2001.
- Goodin, R.E. *The Politics of Rational Man.* (John Wiley) London, 1976.
- Goodin, R.E. *Green Political Theory.* (Polity Press) Oxford, 1992.
- Goodsell, C.T. *The Case for Bureaucracy. A Public Administration Polemic.* Second edition. (Chatham House Publishers) Chatham, 1983/1985.
- Gordon, H.S. 'The Economic Theory of a Common-Property Resource: the Fishery.' In: *Journal of Political Economy* LXII/. (1954) 124-142.
- Gorz, A. *Capitalism, Socialism, Ecology.* (Verso) London, 1994.
- Gould, C.C. *Rethinking Democracy. Freedom and Social Cooperation in Politics, Economy and Society.* (Cambridge University Press) New York, 1988.
- Gouldner, A. *Patterns of Industrial Bureaucracy.* (Free Press) Glencoe, 1954.
- Graaf, H. van de, 'Beleid en beoordeling van beleidstheorieën.' In: *Beleid en Maatschappij* XV/. (1988) 7-21.
- Granovetter, M. and R. Swedberg (eds.) *The Sociology of Economic Life.* (Westview Press) Boulder, 1992a.
- Granovetter, M. 'Economic Action and Social Structure: the problem of Embeddedness.' In: M. Granovetter and R. Swedberg (eds.) *The Sociology of Economic Life.* (Westview Press) Boulder (1992b) 53-81.
- Grant, W. (ed.) *The Political Economy of Corporatism.* (MacMillan) London, 1985.
- Gray, J. *False Dawn. The Delusions of Global Capitalism.* (The New Press) New York, 1998.
- Groenewegen, J. 'De transactiekostentheorie nader bezien. Een toelichting aan de hand van het vraagstuk van de organisatie van arbeid.' In: *Tijdschrift voor Politieke Economie* ./4 (1990) 50-76.
- Groenewegen, P. (ed.) *Economics and Ethics?* (Routledge) London, 1996.
- Guillet de Monthoux, P. *The Moral Philosophy of Management. From Quesney to Keynes.* (M.E. Sharpe) Armonk and New York, 1993.
- Gunsteren, H.R. van *The Quest for Control. A Critique of the Rational-Central-Rule Approach in Public Affairs.* (John Wiley and Sons) London, 1976.
- Gunsteren, H. van, and E. van Ruyven (eds.) *Bestuur in de ongekende samenleving.* (Sdu) The Hague, 1995.
- Haan, I. de, *Zelfbestuur en staatsbeheer. Het politieke debat over burgerschap en rechtsstaat in de twintigste eeuw.* (Amsterdam University Press) Amsterdam, 1993.
- Habermas, J. 'Technik und Wissenschaft als "Ideologie".' In: J. Habermas *Technik und Wissenschaft als "Ideologie".* Frankfurt am Main (1968) 48-118.
- Habermas, J. 'Aspects of the Rationality of Action.' In: T.F. Geraets (ed.) *Rationality Today.* Ottawa, 1979.
- Habermas, J. (ed.) *Stichworte zur 'geistigen Situation der Zeit' Band 1: Nation und Republik.* (Suhrkamp) Frankfurt am Main, 1979.
- Habermas, J. *Theorie des kommunikativen Handelns.* Two Volumes. (Suhrkamp) Frankfurt am Main, 1981.
- Habermas, J., (W. van der Burg and W. van Reijen (eds.)) *Recht en moraal. Twee voordrachten.* (Kok Agora) Kampen, 1988.
- Habermas, J. (and M. Korthals (ed.)) *De nieuwe onoverzichtelijkheid en andere opstellen.* (Boom) Meppel, 1989.

- Habermas, J. *Faktizität und Geltung*. (Suhrkamp) Frankfurt am Main, 1992.
- Hafkamp, W. and G. Molenkamp 'Tussen Droom en Daad: Over Uitvoeren en Handhaven.' In: Commissie Lange Termijn Milieubeleid. *Het milieu: Denkbeelden voor de 21ste Eeuw.* (Kerckebosch) Zeist (1990) 209-251.
- Hafkamp, W. and R. Weterings 'Van Conventies naar maatschappelijk ondernemen.' *typoscript 1999.*
- Hamlin, A.P. *Ethics, Economics and the State.* (Wheatheaf Books) Brighton, 1986.
- Hamrin, R.D. *A Renewable Resource Economy.* New York, 1983.
- Handelingen der Eerste Kamer, vergaderjaar 1991-1992 *Gezondheids- en welzijnswet voor Dieren (16447) nr. 88a* (Sdu) The Hague, 1992.
- Handelingen der Tweede Kamer, zitting 1981 *Gezondheidswet voor Dieren (16447) nr. 5.* (Sdu) The Hague, 1981.
- Handelingen de Tweede Kamer, vergaderjaar 1984-1985 *Gezondheidswet voor Dieren (16447) nr. 6.* (Sdu) The Hague, 1985.
- Handelingen der Tweede Kamer, vergaderjaar 1985-1986 *Gezondheids- en Welzijnswet voor Dieren (16447) nr. 8.* (Sdu) The Hague, 1986.
- Handelingen van de Tweede Kamer der Staten Generaal, vergaderjaar 1988-1989 *Nationaal Milieubeleidsplan (21.149) nrs. 2 en 3* (Sdu) The Hague, 1989.
- Handelingen der Tweede Kamer, vergaderjaar 1994-1995 *Dynamiek en Vernieuwing (24.140) nr. 1/2.* (Sdu) The Hague, 1995.
- Harbers, H. 'Politiek van de technologie.' In: *Kennis en Methode* XX/3 (1996) 308-315.
- Harbers, H. and S. Koenis 'Ter inleiding. De bindkracht der dingen.' In: *Kennis en Methode. Tijdschrift voor empirische filosofie* XXIII/2 (1999) 3-9.
- Hardin, G. 'The Tragedy of the Commons.' In: *Nature* CLXII/. (December 1968) 1243-1248.
- Hardin, G. and J. Baden (eds.) *Managing the Commons.* () San Francisco, 1977.
- Hardin, R. *Collective Action. A Book from Resources for the Future.* () Baltimore, 1982.
- Harmsen, G. *Natuur, geschiedenis, filosofie.* (Sun) Nijmegen, 1974.
- Harmsen, G. *Natuurbeleving en arbeidersbeweging. De Nederlandse socialistische arbeidersbeweging in haar relatie tot natuur en milieu.* (Nivon) Amsterdam, 1992.
- Hart, P. 't 'Politieke besluitvorming in Nederland: een decennium van onderzoek in beeld.' In: *Beleidswetenschap* VI/3 (1992) 199-227.
- Hausman, D.M. and M.S. McPherson 'Taking Ethics Seriously: Economics and Contemporary Moral Philosophy.' In: *Journal of Economic Literature* XXXI/. (June 1993) 671-731.
- Hayek, F.A. *Individualism and Economic Order.* (The University of Chicago Press/Midway) Chicago, 1948/1980a.
- Hayek, F.A. 'Individualism: True and False.' In: F.A. Hayek (ed.) *Individualism and Economic Order.* (The University of Chicago Press/Midway) Chicago (1948/1980b) 1-33.
- Hayek, F.A. *Law, Legislation and Liberty. A New Statement of the Liberal Principles of Justice and Political Economy. Three Volumes.* (Routledge and Kegan Paul) London and Henley, 1976.
- Hayek, F.A. *The Constitution of Liberty.* (Routledge) London, 1960/1990.
- Heilbroner R.L. *De ontwikkeling van de economische samenleving.* (Spectrum) Utrecht, 1962/1977.
- Heilbroner R.L. *An Inquiry into the Human Prospect.* (Calder and Boyars) London, 1975.
- Heilbroner R.L. *Twenty-First Century Capitalism.* (UCL Press Limited) London, 1993.
- Held, D. et al. (eds.) *States and Societies.* (Martin Robertson) Oxford, 1983.
- Held, D. and C. Pollitt (eds.) *New Forms of Democracy.* (Sage/The Open University) London, 1986.
- Held, D. *Models of Democracy.* (Polity Press) Cambridge, 1987.
- Hemerijck, A.C. 'De politiek van de economie.' In: *Beleid en Maatschappij* XXI/5 (September/October 1994) 229-260.
- Hemerijck, A.C. and M. Verhagen 'De institutionele factor: de maatschappelijke inbedding van de economie.' In: *Beleid en Maatschappij* XXI/5 (September/October 1994) 214-216.
- Herfindahl, O.C. and A.V. Kneese *Economic Theory of Natural Resources.* (Charles E. Merrill Publishing Company) Columbus, 1974.
- Hirsch, F. *The Social Limits to Growth.* (Routledge and Kagan Paul) London, 1977.
- Hirschman, A.O. *Exit, Voice and Loyalty. Responses to Decline in Firms, Organizations, and States.* (Harvard University Press) Cambridge Mass., 1970.
- Hirschman, A.O. *The Passions and the Interests. Political Arguments for Capitalism before its Triumph.* (Princeton University Press) Princeton, 1977.

- Hirschman, A.O. 'Rival Interpretations of Market Society: Civilizing, Destructive, or Feeble?' In: *Journal of Economic Literature* XX/. (December 1982) 1463-1484.
- Hirschman, A.O. (ed.) *Rival Views of Market Society and Other Recent Essays.* (Harvard University Press) Cambridge Mass., 1992.
- Hirschman, A.O. *Denken gegen die Zukunft. Die Rhetoriek der Reaktion.* (Carl Hanser Verlag) München, 1991/1992.
- Hobbes, T. (and R. Tuck (ed.)) *Leviathan.* (Cambridge University Press) Cambridge, 1651/1991.
- Hodgson, G.M. *Economics and Institutions. A Manifesto for a Modern Institutional Economics.* (Polity Press) Cambridge, 1988.
- Hoed, P. den, 'Visies op de besturing in de verzorgingsstaat: eenheid en verscheidenheid.' In: *Beleid en Maatschappij* XI/. (1979) 319-329.
- Hoefnagels, E. 'De handel in visvangstrechten in Nederland - De pragmatische maar primitieve werking van een milieurecht.' In: *Beleid en Maatschappij* VI/. (1993) 287-294.
- Hoefnagels, E. 'Kettingreacties in het visserijbeleid. Via onbedoelde gevolgen naar beoogde beleidseffecten.' In: *Facta* I/4 (1993) 2-6.
- Hoeven, E. van der, 'De ethiek van de beleidsambtenaar.' In: *Filosofie en Praktijk* X/4 (1989) 201-214.
- Holthoon, F. van, 'De Geschiedenis van het Publiek Domein.' In: A. Kreukels and J. Simonis (eds.) *Publiek Domein. De Veranderende Balans tussen Staat en Samenleving. Beleid en Maatschappij Jaarboek 1987-1988.* (Boom) Meppel, 1988.
- Homann, K. 'Marktwirtschaft und Unternehmensethik.' In: S. Blasche, W. Köhler and P. Rohs (eds.) *Markt und Moral. Die Diskussion um die Unternehmensethik.* (Haupt) Bern (1994) 109-130.
- Honneth, A. 'Soziologie. Eine Kolumne. Konzeptionen der "civil society".' In: *Merkur. Deutsche Zeitschrift für europäisches Denken* XLVI/1 (1992) 61-65.
- Hood, C.C. *The Limits of Administration.* () London and New York, 1976.
- Hoogendijk, W. *The Economic Revolution.* (International Books) Utrecht, 1991.
- Hoogendijk, W. *Economie ondersteboven.* (Jan van Arkel) Utrecht, 1993.
- Hoogerwerf, A. (ed.) *Overheidsbeleid.* (Samsom) Alphen aan den Rijn, 1978.
- Hoogerwerf, A. (ed.) *Succes en falen van overheidsbeleid in Nederland.* (Samsom) Alphen aan den Rijn, 1983.
- Hoppe, R. 'Voor moed, beleid en... betrouwbare beleidstheorie? Een tussentijdse bezinning.' In: *Beleid en Maatschappij* XV/. (1988) 55-65.
- Hösle, V. 'Economie en Ecologie.' In: R. van Schomberg (ed.) *Het Discursieve Tegengif.* (Kok-Agora) Kampen, 1996.
- Hotelling, H. 'The Economics of Natural Resources.' In: *Journal of Political Economy* IXXXX/2 (April 1939) 137-175.
- Hughes, T. *Networks of Power. Electrification in Western Society, 1880-1930.* (John Hopkins University Press) Baltimore, 1983.
- Huntingdon, S. 'Post-Industrial Politics: How Benign will it be?' *Comparative Politics.* VI/. (1975) 163-192.
- Hutten, Th. and H. Rutten 'De druk der omstandigheden; technologische trajecten in de Nederlandse landbouw.' In: A.L.G.M. Bauwens, M.N. de Groot and K.J. Poppe (eds.) *Agrarisch bestaan. Beschouwingen bij vijftig jaar Landbouw Economisch Instituut.* (Van Gorcum) Assen (1990) 125-145.
- Hyneman, C.S. *Bureaucracy in a Democracy.* (Harper & Brothers) New York, 1950.
- Ierland, E.C. van, 'Environmental Quality as a Target of Economic Policy.' Paper gepresenteerd op *First Congress of the International Society for Intercommunication of New Ideas.* (University of Paris, Sorbonne IV) Paris, 1990.
- Ingram, D. 'The Limits and Possibilities of Communicative Ethics for Democratic Theory.' In: *Political Theory* XXI/2 (May 1993) 294-321.
- Jacobs, M. *The Green Economy. Environment, Sustainable Development and the Politics of the Future.* () Boulder and London, 1991.
- Jänicke, M. 'Democracy as a Condition for Environmental Policy Success: the Importance of Non-Institutional Factors.' In: W. Lafferty and J. Meadowcroft (eds.) *Democracy and the Environment.* (Edward Elgar) Cheltenham, 1997: 71-85.
- Janoski, T. *Citizenship and Civil Sociery. A Framework of Rights and Obligations in Liberal, Traditional and Social Democratic Regimes.* (Cambridge University Press) Cambridge, 1998.
- Jentoft, S. 'Fisheries Co-Management, Delegating Government Responsibility to Fishermen's Organisations.' In: *Marine Policy* XIII/2 (April 1989) 137-154.

- Jentoft, S. and B. McCay 'Usergroup Participation in Fisheries Management: Lessons Drawn from International Experiences.' In: *Marine Policy* XIX/3 (1995) 227-246.
- Jentoft, S., B.J. McCay and D.C. Wilson 'Social Theory and Fisheries co-management.' In: *Marine Policy* XXII/4-5 (1998) 423-436.
- Jentoft, S. and K.N. Mikalsen 'From User Groups to Stakeholders? The Public Interest in Fisheries Management.' In: *Marine Policy* XXV/. (2001) 281-292.
- Jong, H.W. de, *Dynamische markttheorie. Bedrijfseconomische monografieën 49*. (Stenfert Kroese) Leiden, 1985.
- Kalma, P. *De illusie van de democratische staat. Kanttekeningen bij het sociaal-democratisch staats- en democratiebegrip*. Deventer, 1983.
- Katz, W.G. 'Responsibility and the Modern Corporation.' In: *Journal of Law and Economics* III/. (October, 1960) 75- 85.
- Keane, J. *Public Life in Late Capitalism. Toward a Socialist Theory of Democracy*. (Cambridge University Press) Cambridge, 1984.
- Keane, J. *Democracy and Civil Society. On the Predicament of European Socialism, the Prospects for Democracy, and the Problem of Controlling Social and Political Power*. (Verso) London, 1988.
- Keane, J. (ed.) *Civil Society and the State*. (Verso) London, 1988.
- Keane, J. 'The Limits of State Action.' In: J. Keane (ed.) *Democracy and Civil Society*. (Verso) London, 1988c: 1-30.
- Keller, B. 'Olsons Logik des kollektiven Handelns. Entwicklung, Kritik und eine Alternative.' In: *Politische Vierteljahresschrift* XXIX/3 (1988) 388-406.
- Kesting, S. 'Dialog and Discourse Within and Among Communities.' *Paper Presented at the 13th Annual Meeting of the SASE*. Amsterdam, 30 June 2001.
- Kettner, M. 'Rentabilität und Moralität. Offene Probleme in Karl Homanns Wirtschaft- und Unternehmensethik.' In: S. Blasche and W. R. Köhler et al. (eds.) *Markt und Moral. Die Diskussion um die Unternehmensethik*. (St. Galler Beiträge zur Wirtschaftsethik, no 13.) (Haupt / Forum für Philosophie Bad Homburg) Bern (1994) 241-267.
- Keulartz, J. and M. Korthals (eds.) *Museum Aarde. Natuur: criterium of constructie?* (Boom) Amsterdam, 1997.
- Keynes, J.M. *The End of Laissez-Faire*. (Leonard & Virginia Woolf) London, 1926.
- Kickert, W.J.M. *Overheidsplanning. Theorieën, technieken en beperkingen*. (Van Gorcum) Assen, 1986.
- Klijn, E.H. and G. Teisman 'Besluitvorming in beleidsnetwerken: een theoretische beschouwing over het analyseren en verbeteren van beleidsprocessen in complexe beleidsstelsels.' In: *Beleidswetenschap* VI/1 (1992) 32-51.
- Kneese, A.V. *Economics and the Environment*. (Penguin Books) Harmondsworth, 1977.
- Knight, F.H. 'Ethics and Economic Interpretation.' In: F. H. Knight *The Ethics of Competition and Other Essays*. (George Allen & Unwin) London (1935/1951a) 19-40.
- Knight, F.H. 'The Ethics of Competition.' In: F. H. Knight *The Ethics of Competition and Other Essays*. (George Allen & Unwin) London (1935/1951b) 41-75.
- Knight, F.H. 'Value and Price.' In: F. H. Knight *The Ethics of Competition and Other Essays*. (George Allen & Unwin) London (1935/1951c) 237-250.
- Knight, F.H. 'Marginal Utility Analysis.' In: F. H. Knight *The Ethics of Competition and Other Essays*. (George Allen & Unwin) London (1935/1951d) 148-161.
- Knight, F.H. 'Fallacies in the Interpretation of Social Cost.' In: F. H. Knight *The Ethics of Competition and Other Essays*. (George Allen & Unwin) London (1935/1951e) 217-237.
- Knoester, A. *Economische politiek in Nederland*. (Stenfert Kroese) Leiden, 1989.
- Kockelkoren, P. 'De muis in de klauwen van de kat: een kader voor een hermeneutiek van de natuur.' *typoscript*.
- Kooiman, J., S.J. Eldersveld and Th. van der Tak *Bestuur en beleid. Politiek en bestuur in de ogen van kamerleden en hoge ambtenaren. (*Van Gorcum) Assen, 1980.
- Kooiman, J. *Besturen. Maatschappij en overheid in wisselwerking*. (Van Gorcum) Assen, 1988.
- Kooiman, J. and M. van Vliet 'Governance and Public Management.' In: K. Eliassen and J. Kooiman (eds.) *Managing Public Organizations. Lessons from Contemporary European Experiences*. (Sage) London (1993a) 58-72.
- Kooiman, J. (ed.) *Modern Governance. New Government - Society Interactions* (Sage) London, 1993b.

- Kooiman, J. 'Governance and Governability: Using Complexity, Dynamics and Diversity.' In: J. Kooiman (ed.) *Modern Governance. New Government - Society Interactions.* (Sage) London (1993c) 35-50.
- Kooiman, J. 'Socio-Political Governance: Introduction.' In: J. Kooiman (ed.) *Modern Governance. New Government - Society Interactions.* (Sage) London (1993d) 1-6.
- Kooiman, J., S. Jentoft and M. van Vliet *Creative Governance. Opportunities for Fisheries in Europe.* (Ashgate) Aldershot, 1999.
- Korthals, M. 'Duurzaamheid en democratie.' Rede uitgesproken voor Rotaryclub Bussum *typoscript,* 1994.
- Korthals, M. 'Van welvaartsstaat naar preventiestaat (of: tegen milieubeleid met de middelen van gisteren).' *typoscript.*
- Korthals, M. *Duurzaamheid en democratie.* (LUW) Wageningen, 1994.
- Koslowski, P. *Ethik des Kapitalismus.* (J.C.B. Mohr/Paul Siebeck) Tübingen, 1986.
- Kreukels, A.M.J. and J.B.D. Simonis (eds.) *Publiek domein. De veranderende balans tussen staat en samenleving. Beleid en maatschappij jaarboek 1987-1988.* (Boom) Meppel, 1988.
- Kuhn, T. *The Essential Tension.* Chicago, 1977.
- Kuipers, T.A.F. (ed.) *Filosofen in actie.* (Eburon) Groningen, 1992.
- Kunneman, H. *Habermas' theorie van het communicatieve handelen. Een samenvatting.* (Boom) Meppel, 1983.
- Kunneman, H. 'Democratie, milieu en verlangen: aanzet tot een post-moderne milieupolitiek.' In: *Oikos* I/1 (Autumn 1996) 39-53.
- Kurer, O. 'John Stuart Mill on Government Intervention.' In: *History of Political Thought* X/3 (August 1989) 457-480.
- Kuypers, P. 'Een nieuw corporatisme.' *typoscript.*
- Kwast, C.A., R. Klos, P. Verwey and G.M. van Asperen 'Verzorgingsstaat ter discussie.' In: *Filosofie en Praktijk* III/2 (1982) 70-90.
- Kymlicka, W. *Contemporary Political Philosophy. An Introduction.* (Clarendon Press) Oxford, 1990/1995.
- Lafferty, W.M. and J. Meadowcroft (eds.) *Democracy and the Environment. Problems and Prospects.* (Edward Elgar) Cheltenham, 1997a.
- Lafferty, W. and J. Meadowcroft 'Democracy and the Environment: Prospects for Greater Congruence.' In: W. Lafferty and J. Meadowcroft (eds.) *Democracy and the Environment.* (Edward Elgar) Cheltenham (1997b) 256-271.
- Lash, S. and J. Urry *The End of Organized Capitalism.* (Polity Press) Cambridge, 1987.
- Law, J. (ed.) *A Sociology of Monsters: Essays on Power, Technology and Domination.* (Routledge) London, 1991.
- Leeuw, F.L. and H. van de Graaf 'Onderzoek naar beleidstheorieën: uitgangspunten, methodische aspecten en praktische relevantie.' In: *Beleid en Maatschappij* XV/1 (1988) 1-5.
- Lemstra, W. *Handboek overheidsmanagement.* (Samsom) Alphen aan den Rijn, 1988.
- Liagre Böhl, H. de, et al. *Nederland industrialiseert! Politieke en ideologische strijd rondom het naoorlogse industrialisatiebeleid 1945-1955.* (SUN) Nijmegen, 1981.
- Liebrand, W.B.G. and P.A.M. van Lange *Als het mij maar niets kost! De psychologie van sociale dilemma's.* Amsterdam, 1990.
- Lieshout, R.H. 'Staat en onderontwikkeling. Een kritiek op het constructivistisch-rationalisme.' In: *Beleid en Maatschappij* XV/3 (1988) 145-155.
- Lindblom, C.E. *Politics and Markets. The World's Political-Economic Systems.* (Basic Books) New York, 1977.
- Lindblom, C.E. 'The Accountability of Private Enterprise: Private-No. Enterprise-Yes.' In: T. Tinker (ed.) *Social Accounting for Corporations: Private Enterprise versus the Public Interest.* (Markus Wieners Publishing) New York (1984) 13-36.
- Lipsky, M. *Street-Level Democracy. Dilemma's of the Individual in Public Service.* (Russell Sage Foundation) New York, 1980.
- Lively, J. *Democracy.* Bristol, 1975.
- Locke, J. (and P. Laslett (ed.)) *Two Treatises of Government.* (Cambridge University Press) Cambridge, 1688/1997.
- Long, N.E. 'Power and Administration.' In: F.A. Rourke (ed.) *Bureaucratic Power in National Politics.* (Little, Brown and Company) Boston (1965) 14-22.

- Luijk, H. van, *Aantekeningen voor een bedrijfsethiek*. (Eburon) Delft, 1985.
- Luijk, H. van, *Om Redelijk Gewin. Oefeningen in Bedrijfsethiek*. (Boom) Amsterdam, 1993a.
- Luijk, H. van, 'Rechten en belangen in een participatieve marktsamenleving.' In: H. van Luijk *Om redelijk gewin. Oefeningen in bedrijfsethiek*. (Boom) Amsterdam (1993b) 190-218.
- Luijk, H. van, 'Rights and Interests in a Participatory Market Society.' In: *Business Ethics Quarterly* IV/1 (1994) 79-96.
- Luijk, H. van, 'Multinationale Ondernemingen Zouden Moeten... .' In: *Filosofie en bedrijf*. ./33 (1999a) 16-27.
- Luijk, H. van, 'Business Ethics in Europe: a Tale of Two Efforts.' In: R.E. Frederick (ed.) *A Companion to Business Ethics*. (Blackwell Companions to Philosophy) Malden (1999b) 353-365.
- Luijk, H. van, *Concurrentie en de Moraal van de Markt*. (Universiteit van Nijenrode) Breukelen, 2000.
- Luijk, H. van, 'Integriteit in het publieke domein. Contouren van een representatieve ethiek.' *typoscript*, Oktober 2001.
- Lukes, S. 'The New Democracy.' In: S. Lukes *Essays in Social Theory*. () London (1977) 30-51.
- Maanen, G.E. van *Publiek domein en het belang van de overheid bij bodemsanering. Een ongewasschen varken?* (Kluwer) Deventer, 1990.
- MacIntyre, A. *After Virtue: A study in Moral Theory*. (Duckworth) London, 1981.
- MacLagan, P. *Management and Morality. A developmental Perspective*. (Sage) London, 1998.
- MacPherson, C.B. *Democratic Theory. Essays in Retrieval*. (Clarendon Press) Oxford, 1973/1989.
- MacPherson, C.B. 'Berlin's Division of Liberty.' In: C.B. MacPherson (ed.) *Democratic Theory. Essays in Retrieval*. (Clarendon Press) Oxford (1973/1989b) 95-119.
- MacPherson, C.B. *The Life and Times of Liberal Democracy*. (Oxford University Press) Oxford, 1977.
- Mahoney, J. and E. Vallance *Business Ethics in a New Europe*. (Kluwer) Dordrecht, 1992.
- Mair, D. and A.G. Miller (eds.) *A Modern Guide to Economic Thought. An Introduction to Comparative Schools of Thought in Economics*. (Edward Elgar) Brookfield, 1991.
- Maitland, I. 'The Limits of Business Self-Regulation.' In: *California Management Review* XXVII/. (Spring 1985) 132-147.
- Majone, G. *Evidence, Argument and Persuasion in the Policy Process. Changing Institutional Constraints*. London, 1989.
- Mandeville, B. (and F.B. Kaye (ed.)) *The Fable of the Bees or Private Vices, Publick Benefits. Two Volumes*. (Clarendon Press) Oxford, 1714/1924/1966.
- Mann, D.E. (ed.) *Environmental Policy Implementation. Planning and Management Options and their Consequences*. (Lexington Books) Lexington, 1982.
- Mannheim, K. *Freedom, Power & Democratic Planning*. (Routledge and Kegan Paul) London, 1950/1968.
- Mansbridge, J.J. (ed.) *Beyond Selfinterest*. (University of Chicago Press) Chicago, 1990.
- Marcet Mrs., *Conversations on Political Economy: In Which The Elements of That Science Are Famililarly Explained*. () London, 1816/1839.
- Marcus, A. 'Converting Thought into Action: The Use of Economic Incentives to Reduce Pollution.' In: D. Mann (ed.) *Environmental Policy Implementation*. (Lexington Books) Lexington, 1982: 173-183.
- Marshall A, *Principles of Economics. An Introductory Volume*. Eighth edition. (Porcupine Press) Philadelphia/Pennsylvania, 1890/1990.
- Marx, K. *Das Kapital. Kritik der Politischen Ökonomie. Volume I. (Marx-Engels-Werke Volume 23)*. (Dietz Verlag) Berlin, 1867/ 1970.
- Mashaw, J.L. *Due Process in the Administrative State*. (Yale University Press) New Haven, 1985.
- Maus, I. 'Verrechtlichung, Entrechtlichung und der Funktionswandel von Institutionen.' In: I. Maus *Rechtstheorie und politische Theorie im Industriekapitalismus*. (Fink Verlag) München, 1986.
- Mayntz, R. et al. *Vollzugsprobleme der Umweltpolitik. Empirische Untersuchung der Implementation von Gesetzen im Bereich der Luftinhaltung und des Gewässerschutzes*. (Kohlhammer) Stuttgart, 1978.
- Mayntz, R. 'Governing Failures and the Problem of Governability.' In: J. Kooiman (ed.) *Modern Governance. New Government - Society Interactions*. (Sage) London (1993) 9-20.
- Mayntz, R. 'Nieuwe uitdagingen voor de governance theorie.' In: *Beleid en maatschappij* XXVI/1 (1999) 2-12.
- McCay, B.J. and J.M. Acheson *The Question of the Commons: the Culture and Ecology of Communal Resources*. () Tucson, 1990.
- McCay, B. and S. Jentoft 'Uncommon Ground: Critical Perspectives on Common Property.' In: *Kölner Zeitschrift für Soziologie und Sozialpsychologie* ./. (1996)

- McComas, M. 'Into the Political Arena.' In: J. Dobbing (ed.) *Infant Feeding. Anatomy of a Controversy 1973-1984.* (Springer Verlag) Berlin, (1988) 63-81.
- McKie, J.W. (ed.) *Social Responsibility and the Business Predicament.* (The Brookings Institution) Washington D.C., 1974.
- Meadows, D.H., D.L. Meadows, J. Randers et al. *The Limits to Growth. A Report for the Club of Rome's Project on the Predicament of Mankind.* (Earth Island) London, 1972.
- Mentzel, M. *Milieubeleid. Normatief bezien.* (Stenfert Kroese) Leiden, 1993.
- Merton, R.K., A.P. Gray, B. Hockey and H.C. Selvin (eds.) *Reader in Bureaucracy.* (The Free Press/MacMillan Publishing Company) New York, 1952.
- Michalski, K. (ed.) *Europa und die Civil Society.* (Klett-Cotta) Stuttgart, 1991.
- Michiels, F.C.M.A. *De wet milieubeheer.* (Tjeenk Willink) Zwolle, 1992.
- Mill, J.S. (and D. Winch (ed.)) *Principles of Political Economy. With Some of Their Applications to Social Philosophy. Books IV and V.* (Penguin Books) London, 1848/1970.
- Mill, J.S. (and G. Himmelfarb (ed.)) *On Liberty.* (Penguin Books) London, 1859/1985.
- Ministerie van Volksgezondheid, Ruimtelijke Ordening en Milieubeheer *Tweede Nationaal Milieubeleidsplan.* (Sdu) The Hague, 1993.
- Ministerie van Volksgezondheid, Ruimtelijke Ordening en Milieubeheer *Nationaal Milieubeleidsplan 3.* (Sdu) The Hague, 1998.
- Mishan, E.J. *Introduction to Normative Economics.* (Oxford University Press) Oxford, 1981.
- Mishan, E.J. *Introduction to Political Economy.* (Hutchinson & Co) London, 1982.
- Mol, A.P.J., G. Spaargaren and A. Klapwijk (eds.) *Technologie en milieubeheer. Tussen sanering en ecologische modernisering.* (Sdu) The Hague, 1991a.
- Mol, A. and G. Spaargaren 'Introductie: Technologie, Milieubeheer en Maatschappelijke Verandering.' In: A. Mol, G. Spaargaren and A. Klapwijk (eds.) *Technologie en Milieubeheer. Tussen Sanering en Ecologische Modernisering.* (Sdu) The Hague (1991b) 5-21.
- Monroe, A.E. (ed.) *Early Economic Thought.* (Harvard University Press) Cambridge, 1924.
- Morgan, G. *Images of Organization.* (Sage) London, 1986.
- Mouzelis, N.P. *Organization and Bureaucracy. An Analysis of Modern Theories.* (Routledge and Kegan Paul) London, 1967/1975.
- Mulberg, J. *The Social Limits of Economic Theory.* () 1995.
- Münckler, H. (ed.) *Die Chancen der Demokratie. Grundprobleme der Demokratie.* (Piper) München, 1991.
- Nagelkerke, A.G. 'De klassieke institutionele economie en de mainstream.' In: *Beleid en Maatschappij* XXI/. (September/October 1994-5) 217-228.
- Nash, R.F. *The Rights of Nature. A History of Environmental Ethics.* (University of Wisconsin Press) Madison, 1989.
- Nauta, L.W. 'Historical Roots of the Concept of Autonomy in Western Philosophy.' In: *Praxis* IV/4 (January 1985) 363-377.
- Nauta, L.W. 'U luistert naar Radio Moskou.' In: *Nieuwe Wereld Tijdschrift* VI/4 (1989) 22-31.
- Nauta, L.W. 'Changing Conceptions of Citizenship.' In: *Praxis International* XII/1 (1992) 20-34.
- Nauta, L.W. 'Plessners Antropologie im gesellschaftlichen Kontext (über die Pragmatik eines Konzepts der menschlichen Person).' In: *Acta Philosophica Groningana. Research Bulletin of the Centre for Philosophical Research ./8,* Groningen, 1993.
- Nauta, L.W. 'Het bolwerk der weerbaren.' typoscript, 1998.
- Nauta, L.W. 'Waar is democratie goed voor?' In: L.W. Nauta (ed.) *Onbehagen in de filosofie.* (Van Gennep) Amsterdam, 2000.
- Nauta, L.W. 'The Third Dimension of Citizenship.' *Typoscript,* 1999.
- Neher, P.A. *Natural Resource Economics. Conservation and Exploitation.* () Cambridge and New York, 1990.
- Nelissen, N.J.M., J. Geurts and H. de Wit (eds.) *Het verkennen van beleidsproblemen.* (Kerckebosch) Zeist, 1986.
- Nelissen, N.J.M. *Besturen binnen verschuivende grenzen.* (Kerckebosch) Zeist, 1992.
- Nelson, R.R. and S.G. Winter *An Evolutionary Theory of Economic Change.* (Harvard University Press) Cambridge Mass., 1982.
- Newton, L.H. 'A passport for the corporate code: from Borg Warner to the Caux Principles.' In: R.E. Frederick (ed.) *A Companion to Business Ethics.* (Blackwell Companions to Philosophy) Malden (1999) 374-385.

- Niskanen, W.A. *Bureaucracy and Representative Government.* () Chicago, 1971.
- North, D.C. and R.P. Thomas *The Rise of the Western World: a New Economic History.* (Cambridge University Press) Cambridge, 1973/1981.
- Norton, G.A. *Resource Economics.* (Edward Arnold) London, 1984.
- Nozick, R. *Anarchy, State, and Utopia.* (Basil Blackwell) Oxford, 1974.
- Offe, C. *Strukturprobleme des kapitalistischen Staates. Aufsätze zur politischen Soziologie.* (Suhrkamp) Frankfurt am Main, 1975.
- Offe, C. '>Unregierbarkeit<' In: J. Habermas *Stichworte zur 'geistigen Situation der Zeit'. Band I.* (Suhrkamp) Frankfurt aM. (1979) 294-318.
- Offe, C. 'The Divergent Rationalities of Administrative Action.' In: C. Offe and J. Keane (eds.) *Disorganized Capitalism. Contemporary Transformations of Work and Politics.* (Polity Press) Cambridge (1985) 300-316.
- Olson, M. *The Logic of Collective Action. Public Goods and the Theory of Groups.* (Harvard University Press) Cambridge Mass., 1965.
- O'Neill, J.J. (ed.) *Modes of Individualism and Collectivism.* (Heinemann) London, 1973.
- Ophuls, W. *Ecology and the Politics of Scarcity. Prologue to a Political Theory of the Steady State.* (W.H. Freeman) San Francisco, 1977.
- Opschoor, H. *Na ons geen zondvloed. Voorwaarden voor een duurzaam milieugebruik.* (Kok Agora) Kampen, 1989.
- Ordeshook, P.C. *Game Theory and Political Theory. An Introduction.* () Cambridge Mass., 1986.
- O'Riordan, T. 'Democracy and the Sustainable Transition.' In: W. Lafferty and J. Meadowcroft (eds.) *Democracy and the Environment.* (Edward Elgar) Cheltenham 1997) 140-156.
- Ostrom, E. *Governing the Commons. The Evolution of Institutions for Collective Action.* (Cambridge University Press) New York, 1990.
- Ouchi, W. 'Markets, Bureaucracies and Clans.' In: *Administrative Science Quarterly XXV/.* (1980) 129-141.
- Paehlke, R.C. *Environmentalism and the Future of Progressive Politics.* (Yale University Press) New Haven, 1989.
- Paine, L.S. *(Cases in) Leadership, Ethics and Organizational Integrity. A Strategic Perspective.* (Irwin) Chicago, 1997.
- Paine, L.S. 'Does Ethics Pay?' In: *Business Ethics Quarterly X/1* (January 2000) 319-330.
- Parnell, M.F. *The German Tradition of Organized Capitalism. Self-Government in the Coal Industry* Oxford, 1994.
- Parsons, T. and N. Smelzer *Economy and Society. A Study in the Integration of Economic and Social Theory.* (Routledge and Kegan Paul) London, 1957.
- Pateman, C. *The Problem of Political Obligation. A Critique of Liberal Theory.* (Polity Press) Oxford, 1979/1985.
- Pateman, C. 'Social Choice or Democracy? A Comment on Coleman and Ferejohn.' In: *Ethics ./97* (October 1986) 39-46.
- Paul, E.F., F.D. Miller jr. and J. Paul *Ethics and Economics.* (Basil Blackwell) Oxford, 1985.
- Pearce, D.W. and R.K. Turner *Economics of Natural Resources and the Environment.* (Harvester Wheatsheaf) New York, 1990.
- Pearce, D.W., R.K. Turner and I. Bateman *Environmental Economics. An Elementary Introduction.* (Harvester Wheatsheaf) New York, 1994.
- Pearse, P.H. 'Fishing Rights and Fishing Policy: the Development of Property Rights as Instruments of Fisheries Management.' In: Vogtlandes (ed.) *The State of the World's Fisheries Resources. Proceedings of the World Fisheries Congress Planning Session.* (IBH Publishing Company) Oxford, 1994.
- Pels, D. 'De "natuurlijke saamhorigheid" van feiten en waarden.' In: *Kennis en Methode XIV/.* (1990) 14-43.
- Pels, D. 'Elster's Tocqueville. Enkele Kritische Kanttekeningen.' In: T. Kuipers (ed.) *Filosofen in Actie.* (Eburon) Groningen (1992) 109-118.
- Pels, D. *Het democratisch verschil. Jacques de Kadt en de nieuwe elite.* (Van Gennep) Amsterdam, 1993.
- Pepperman Taylor, B. 'Democracy and Environmental Ethics.' In: W. Lafferty and J. Meadowcroft (eds.) *Democracy and the Environment.* (Edward Elgar) Cheltenham, 1997: 86-107.
- Perrow, C. *Complex Organizations. A Critical Essay.* (Scott, Foresman and Company) Glenview, 1972.

- Perrow, C. 'Normal Accidents at Three Mile Island.' In: C. Perrow *Normal Accidents. Living with High-Risk Technologies.* (Basic Books) New York, 1984.
- Pesch, H. s.j., *Ethiek en volkshuishoudkunde.* () Eindhoven, 1928.
- Peters, A.A.G. 'Recht als project.' In: *Ars Aequi* XVIII/11 (1979) 245-256.
- Peterse, A.H. *Onderneming en politiek in liberale democratieën.* (Wolters-Noordhoff) Groningen, 1990.
- Pigou, A.C. *The Economics of Welfare.* Fourth edition. (MacMillan) New York, 1920/1962.
- Pinkerton, E. *Co-Operative Management of Local Fisheries. New Directions for Improved Management and Community Development.* () Vancouver, 1989.
- Polanyi, K. *The Great Transformation.* (Rinehart and Company) New York, 1944.
- Porter, M.E. *Porter over concurrentie.* (Uitgeverij Contact) Amsterdam, 1999.
- Porter, M.E. *Concurrentievoordeel. De beste resultaten behalen en behouden.* (Uitgeverij Contact) Amsterdam, 1999.
- Pot, . van der, A.M. Donner et al. *Handboek van het Nederlandse staatsrecht.* Thirteenth edition. (Tjeenk Willink) Zwolle, 1940/1995.
- Powel, W.W. 'Neither Market nor Hierarchy: Network Forms of Organization.' In: *Research in Organizational Behavior* XII/. (1990) 295-336.
- Pressman, J.L. and A.B. Wildavski *Implementation. How Great Expectations in Washington Are Dashed in Oakland.* (University of California Press) Berkeley, 1973.
- Preston, L.E. and J.E. Post *Private Management and Public Policy. The Principle of Public Responsibility.* (Prentice Hall) Englewood Cliffs, 1975.
- Pribram, K. *A History of Economic Reasoning.* (John Hopkins University Press) Baltimore, 1983.
- Pritwitz, V. 'Multifacetted Analysis of International Environmental Policy.' In: *Industrial Crisis Quarterly* ./3 (1989) 77-99.
- Putnam, R., (R. Leonardi and R.Y. Nanetti) *Making Democracy Work. Civic Traditions in Modern Italy.* (Princeton University Press) Princeton, New Jersey, 1993.
- Rainey, H.G. and H.B. Milward 'Public Networks and the Environment.' In: R.H. Hall and R.E. Quinn (ed.) *Organizational Theory and Public Policy.* Beverly Hills, 1983.
- Rawls, J. *A Theory of Justice.* (Oxford University Press) Oxford, 1972.
- Reijnders, L. *Pleidooi voor een duurzame relatie met het milieu.* (Van Gennep) Amsterdam, 1984.
- Reijnders, L. *Het milieu, de politiek en de drie verkiezingen.* (Van Gennep) Amsterdam, 1993.
- Reisman, D. *Theories of Collective Action. Downs, Olson, Hirsch.* (MacMillan) London, 1990.
- Riemsdijk, M.J. *Actie of dialoog. Over de betrekkingen tussen maatschappij en onderneming.* (self published) Amsterdam (1994) 43-61.
- Riemsdijk, M. 'De Brent Sparr als symptoom.' In: *Holland Management Review* no 44 (1995) 82-86.
- Riemsdijk, M. 'Actie, dialoog en onderhandeling.' In: H. Tieleman, H. van Luijk et al. (eds.) *Conflicten tussen Actiegroepen en Ondernemingen.* (Smo) The Hague, 1996: 68-96.
- Righart, H. *Het einde van Nederland? Kenteringen in politiek, cultuur en milieu.* (Uitgeverij Kosmos) Utrecht, 1992.
- Riker, W. and P. Ordeshook *An Introduction into Positive Political Theory.* (Prentice Hall) Englewood Cliffs, 1973.
- Ringeling, A. 'Dynamiek van de Overheidsbemoeienis: Groei van Tal and Last.' In: J. de Beus and J. van Doorn (eds.) *De Interventiestaat. Traditics, Ervaringen, Reacties.* (Boom) Meppel, 1984: 117-137.
- Robertson, D.H. *Economic Commentaries.* (Hyperion Press) Westport, 1956/1979.
- Robinson, J. *Economic Philosophy.* (Penguin Books) Harmondsworth, 1962/1978.
- Rödel, U., G. Frankenberg and H. Dubiel *Die demokratische Frage.* (Suhrkamp) Frankfurt am Main, 1989.
- Roobeek, A.J.M. 'De smalle marges van het technologiebeleid.' In: *Beleid en Maatschappij* XVI/. (1989) 246-257.
- Roorda, F.A. 'De uitvoering van de Grondwaterwet, een instrument voor grondwaterbeheer.' In: P.J.J. Buuren and G. Betlem *Milieurecht in stelling. Utrechtse opstellen over actuele thema's in het milieurecht.* (W.E.J. Tjeenk-Willink) Zwolle (1990) 124-150.
- Rosanvallon, P. 'The Decline of Social Visibility.' In: J. Keane (ed.) *Civil Society and the State.* (Verso) London (1988) 199-220.
- Rose, F. de 'Utilisme en Natuurbehoud.' In: *Filosofie en Praktijk* IX/3 (1988) 128-142.
- Rothschild, K.W. *Ethics and Economic Theory.* (Edward Elgar) Aldershot, 1993.
- Rourke, F.E. (ed.) *Bureaucratic Power in National Politics.* () Boston, 1965/1978.

- Sabine, G.H. and T.L. Thorson *A History of Political Theory.* Fourth edition. (Dryden Press/Holt-Saunders) Hinsdale/Illinois, 1937/1973.
- Sahlins, M. *Stone Age Economics.* (Tavistock Publications) London, 1974/1984.
- Salz, P. *De Europese Atlantische visserij, structuur, economische situatie en beleid.* (LEI-DLO) The Hague, 1991.
- Samuelson, P.A. and W.D. Nordhaus *Economics.* Twelfth edition. (McGraw Hill) New York, 1955/1985.
- Sandel, M. *Liberalism and the Limits of Justice.* (Cambridge University Press) Cambridge, 1982.
- Sartoni, G. *The Theory of Democracy Revisited. Two Volumes.* (Chatham House Publishers) Chatham, 1987.
- Scanlon, T.M. *What We Owe to Each Other.* (The Belknap Press of Harvard University Press) Cambridge Mass., 1998/2000.
- Schelling, T. *Choice and Consequence.* (Harvard University Press) Cambridge Mass., 1984.
- Schelling, T. (ed.) *Incentives for Environmental Protection.* (MIT Press) Cambridge Mass., 1983.
- Schmitter, P.C. and G. Lehmbruch (eds.) *Trends toward Corporatist Intermediation.* (Sage) London, 1979a.
- Schmitter, P. 'Still the Century of Corporatism?' In: P. Schmitter and G. Lehmbruch (eds.) *Trends toward Corporatist Intermediation.* (Sage) London (1979b) 7-51.
- Schmitter, P.C. and G. Lehmbruch *Patterns of Corporatist Policy-Making.* (Sage) London, 1982a.
- Schmitter, P. 'Reflections on Neocorporatism.' In: P. Schmitter and G. Lehmbruch (eds.) *Patterns of Corporatist Policy-Making.* (Sage) London (1982b) 259-279.
- Schmoller, G. *Zur Sozial- und Gewerbepolitik der Gegenwart,* 1890.
- Schomburg, R. van (ed.) *Het discursieve tegengif. De sociale en ethische aspecten van de ecologische crisis.* (Kok-Agora) Kampen, 1997.
- Schotter, A. *Free Market Economics. A Critical Appraisal.* (Basil Blackwell) Cambridge Mass., 1990.
- Schrijvers, P.M.B. and H.C.M. Moor-Smeets *Staats- en Bestuursrecht.* (Wolters-Noordhoff) Groningen, 1991.
- Schultze, C.L. *The Public Use of Private Interest.* (The Brookings Institution) Washington D.C., 1977.
- Schumpeter, J.A. *Capitalism, Socialism and Democracy.* (George Allen & Unwin) London, 1944.
- Schumpeter, J.A. *History of Economic Analysis.* (Routledge) New York, 1954/1986.
- Schuyt, C.J.M. *Recht en samenleving. Centrale problemen, alternatieven en overzichten.* (Van Gorcum) Assen, 1981.
- Schuyt, C.J.M. 'De verzorgingsstaat als object van pars-pro-toto-generalisaties.' In: *Beleid en Maatschappij* XVI/3 (1989) 119-132.
- Schuyt, K. and R. van der Veen (eds.) *De verdeelde samenleving. Een inleiding in de ontwikkeling van de Nederlandse verzorgingsstaat.* (Stenfert Kroese) Leiden, 1990/1992.
- Scott, W.R. *Organizations. Rational, Natural and Open Systems.* Fourth Edition (Prentice Hall International) Upper Saddle River, 1981/1998.
- Scott, W. R. *Institutions and Organizations.* (Sage) Thousand Oaks, 1995.
- Searle, R. *Morality and the Market in Victorian Britain.* (Clarendon Press) Oxford, 1998.
- Selznick, P. *TVA and the Grass Roots.* (University of California Press) Berkeley, 1949.
- Selznick, P. *The Moral Commonwealth. Social Theory and the Promise of Community.* (University of California Press) Berkeley, 1992.
- Sen, A. 'Isolation, Assurance and the Social Rate of Discount.' In: *Quarterly Journal of Economics* LXXXI/. (1967) 112-124.
- Sen, A. 'The Moral Standing of the Market.' In: E. Paul, F. Miller jr. and J. Paul (eds.) *Ethics and Economics.* (Basic Blackwell) Oxford (1985) 1-19.
- Sen, A. *On Ethics and Economics.* (Basil Blackwell) Oxford, 1987.
- Sen, A. 'Money and Value. On the Ethics and Economics of Finance.' In: *Economics and Philosophy.* IX/. (1993) 203-227.
- Sherman, H.J. *The Business Cycle. Growth and Crisis under Capitalism.* (Princeton University Perspective Press) Princeton, 1991.
- Shklar, J.N. *Legalism: Law, Morals and Political Trials.* (Harvard University Press) Cambridge Mass., 1964/1986.
- Simon, J. 'Global Confusion, 1980: a Hard Look at the Global 2000 Report.' In: *Public Interest* (1981) 3-20.

- Simon, J. and H. Kahn (eds.) *The Resourceful Earth: a Response to Global 2000.* (Blackwell) Oxford, 1984.
- Skillen, J.W. *The Development of Calvinistic Political Theory in the Netherlands, with a Special Reference to the Thought of Herman Dooyeweerd.* (University Microfilms International) Ann Arbor, 1974/1979.
- Skinner, Q. *The Foundations of Modern Political Thought. Two Volumes.* (Cambridge University Press) Cambridge, 1978.
- Skolnick, J.H. 'Coercion to Virtue.' In: *Southern California Law Review* XLI/3 (1968) 588-641.
- Smelzer, N. and R. Swedberg (eds.) *The Handbook of Economic Sociology.* (Princeton University Press) New York, 1994.
- Smit, W. 'Visserij-techniek contra een beperkt natuurlijk potentieel.' In: A.L.G.M. Bauwens et al. (eds.) *Agrarisch bestaan. Beschouwingen bij vijftig jaar landbouw-economisch instituut.* (Van Gorcum) Assen (1990) 84-96.
- Smith, A., (R.H. Campbell and A.S. Skinner (eds.)) *An Inquiry into the Nature and Causes of the Wealth of Nations. Two Volumes.* (Oxford University Press) Oxford, 1776/1976a.
- Smith, A. *The Wealth of Nations.* (Everyman's Library) London, 1776/1910/1991b.
- Soroos, M.S. 'Coping with Resource Scarcity: a Critique of Lifeboat Ethics.' In: Kegley and Wittkopf (eds.) *The Global Agenda, Issues and Perspectives.* () New York (1984) 350-366.
- Soroos, M.S. 'Conflict in the Use and Management of International Commons.' In: J. Käkönen *Perspectives on Environmental and International Politics.* () London (1992) 31-43.
- Spoormans, H. *'Met uitsluiting van voorregt'. Het ontstaan van de liberale democratie in Nederland.* () Amsterdam, 1988.
- Stone, C.D. *Where the Law Ends. The Social Control of Corporate Behaviour.* (Harper and Row) New York, 1975.
- Straaten, J. van der, *Zure regen. Economische theorie en het Nederlandse beleid.* (Jan van Arkel) Utrecht, 1990.
- Stuurman, S. *De labyrintische staat. Over politiek, ideologie en moderniteit.* (Sua) Amsterdam, 1985.
- Stuurman, S. *Verzuiling, kapitalisme en patriachaat. Aspecten van de ontwikkeling van de moderne staat in Nederland.* (Sun) Nijmegen, 1983.
- Swaan, A. de *Zorg en de staat. Welzijn, onderwijs en gezondheidszorg in Europa en de Verenigde Staten in de nieuwe tijd.* (Bert Bakker) Amsterdam, 1988/1989.
- Talmon, J.L. *The Origins of Totalitarian Democracy.* () London, 1952.
- Tawney, R.H. *Religie en de opkomst van het kapitalisme.* (Sun) Nijmegen, 1926/1979.
- Taylor, C. *Hegel and Modern Society.* (Cambridge University Press) Cambridge, 1979.
- Taylor, C. 'Die Beschwörung der *Civil Society.*' In: K. Michalski (ed.) *Europa und die Civil Society.* (Klett-Cotta) Stuttgart (1991) 52-81.
- Taylor, M. *The Possibility of Cooperation.* (Cambridge University Press) Cambridge, 1987.
- Taylor, M. and S. Singleton 'The Communal Resource: Transaction Costs and the Solution of Collective Action Problems.' In: *Politics and Society* XXI/2 (June 1993) 195-214.
- Teisman, G.R. *Complexe besluitvorming. Een pluricentrisch perspectief op besluitvorming over ruimtelijke investeringen.* (Vuga) The Hague, 1992.
- Teubner, G. and H. Willke 'Kontext und Autonomie: gesellschaftliche Selbststeuerung durch reflexives Recht.' In: *Zeitschrift für Rechtssoziologie* VI/1 (1984) 4-35.
- Thomassen, J.J.A. (ed.) *Democratie: theorie en praktijk.* (Samsom) Alphen aan den Rijn, 1981.
- Thompson, D.F. 'Moral Responsibility of Public Officials. The problem of many hands.' in: *The American Political Science Review* LXXIV/2 (1980) 905-916.
- Thompson, G., J. Frances et al. *Markets, Hierarchies and Networks. The Coordination of Social Life.* (Sage) London, 1991.
- Thunnissen, F.H.A.M. 'Hoeft rechtszekerheid niet in het milieu?' In: *NJB* VIII/39 (1990) 1511-1516.
- Tieleman, H.J., H.J. van Luijk et al. (eds.) *Conflicten tussen actiegroepen en ondernemingen. De democratisering van het moreel gezag.* (Stichting Maatschappij en Onderneming) The Hague, 1996.
- Tilly, C. 'The Complexity of Popular Collective Action.' In: *New School for Social Research. The Working Paper Series, no. 8* (Center for Studies of Social Change) 1985.
- Tisdell, C. *Environmental Economics. Politics for Environmental Management and Sustainable Development.* (Edward Elgar) Brookfield, 1993.
- Tjeenk Willink, H.D. *Regeren in een dubbelrol.* (Sdu) The Hague, 1980.

- Tjeenk Willink, H.D. *De mythe van het samenhangend overheidsbeleid.* (W.E.J. Tjeenk Willink) Zwolle, 1984.
- Tjeenk Willink, H.D. *De kwaliteit van de overheid. Een bijdrage aan het hernieuwde politieke debat.* (Dop) The Hague, 1989.
- Tocqueville, A. de *Democracy in America. Two Volumes.* New York (z.j.).
- Tufte, E.R. *Political Control over the Economy.* () Princeton, 1978.
- Ulrich, P. *Transformation der Okonomischen Vernunft. Fortschrittperspectiven der modernen Industriegesellschaft.* Third edition. (Haupt) Bern, 1993.
- Ulrich, P. *Integrative Wirtschaftsethik. Grundlagen einer lebensdienlichen Ökonomie.* (Haupt) Bern, 1997.
- Vandevelde, T. 'Van schaarste en overvloed. Een sociaal-filosofische reflectie over de milieucrisis.' In: *Tijdschrift voor Filosofie .*/1 (1992) 16-41.
- Varty, J. 'Civic or Commercial? Adam Ferguson's Concept of Civil Society.' In: Fine, R. *Civil Society. Democratic Perspectives.* (Frank Cass) London 1997) 29-47.
- Ven, B. van de, *Rationaliteit en ethiek in de onderneming.* (Tilburg University Press) Tilburg, 1998.
- Vereniging voor Bestuurskunde *Politisering van het openbaar bestuur.* (Vereniging Nederlandse Gemeenten) The Hague, 1973.
- Verhoog, H. 'Ethiek en milieuproblematiek.' In: *Filosofie en Praktijk* III/1 (1982) 14-30.
- Vermeersch, E. 'Weg van het WTK-complex: onze toekomstige samenleving.' In: Commissie Lange Termijn Milieubeleid *Het milieu: denkbeelden voor de21ste eeuw.* (Kerckebosch) Zeist (1990) 17-44.
- Viner, J. 'The Intellectual History of Laissez Faire.' In: *Journal of Law and Economics.* III/. (October 1960) 45-69.
- Vliet, M. van, *Communicatieve besturing van het milieuhandelen van ondernemingen.* (Eburon) Delft, 1992.
- Vliet, M. van, 'Controlling VOCs by Government-Industry Consensus.' In: *Greener Management International. .*/6 (April 1994) 41-49.
- Vliet, M. van, and W. Dubbink 'Het *Tragedy of the Commons* model en het Nederlandse visserijbeheer.' In: *Beleid en Maatschappij* XXV/1 (1998) 27-39.
- Vliet, M. van, and W. Dubbink 'Evaluating Governance: State, Market and Participation Compared.' In: J. Kooiman, S. Jentoft and M. van Vliet *Creative Governance. Opportunities for Fisheries in Europe.* (Ashgate) Aldershot (1999) 11-32.
- Vogel, D. 'The political and Economic Impact of Current Criticism of Business.' In: *California Management Review* XVIII/2 (winter 1975) 86-92.
- Vogel, D. 'Business Ethics: New Perspectives on Old Problems.' In: *California Management Review.* XXXIII/4 (Summer 1991) 101-117.
- Vries, G. de, 'Leefwereld en systeem - een theorie die mank gaat.' In: *Kennis en Methode* VII/. (1983) 313-328.
- Vroom, B. de, 'Zelfregulering.' In: W. Derksen (ed.) *De terugtred van de regelgevers.* (W.E.J. Tjeenk Willink) Zwolle, 1989.
- Waarden, F.B. van, 'Vervlechting van staat en belangengroepen. Deel II: Wetgeving en zelfregulering.' In: *Beleid en Maatschappij* XV/. (1988) 115-126.
- Waarden, F.B. van, 'Zelfregulering en corporatisme.' In: F.B. van Waarden *Organisatie-macht van belangenverenigingen.*, 1989.
- Wal, K. van der, 'Duurzaamheid: Globalisering en ethiek.' In: M. Becker and K. Klop et al. (eds.) *Economie en Ethiek in Dialoog.* (Van Gorcum) Assen (2001) 97-122.
- Waldo, D. *The Administrative State. A Study of the Political Theory of American Public Administration.* Second edition. (Holmes & Meier) New York and London, 1948/1984.
- Walzer, M. *Spheres of Justice. A Defense of Pluralism and Equality.* (Basic Books) New York, 1983.
- Walzer, M. 'The Idea of Civil Society. A Path to Social Reconstruction.' In: *Dissent* (spring 1991) 293-304.
- Walzer, M. 'The Civil Society Argument.' In: R. Beiner (ed.) *Theorizing Citizenship.* (State University of New York Press) Albany (1995) 153-174.
- Weale, A. *The New Politics of Pollution.* (Manchester University Press) Manchester, 1992.
- Weber, M. *Wirtschaft und Gesellschaft. Grundriss der verstehenden Soziologie.* Fifth edition. Studienausgabe. (J.C.B. Mohr) Tübingen, 1921/1972.
- Weber, M. (and A. van Braam (ed.)) *Gezag en bureaucratie.* (Universitaire Pers Rotterdam) Rotterdam, 1972.

- Weber, P. *Net Loss: Fish, Jobs and Marine Environment.* Worldwatch paper 120, 1994.
- Werhane, P.H. 'Exporting Mental Models: Global Capitalism in the 21st Century.' In: *Business Ethics Quarterly* X/1 (January 2000) 353-362.
- Wetenschappelijke Raad voor het Regeringsbeleid *Milieubeleid. Strategie, instrumenten en handhaafbaarheid; nr. 41 uit de reeks 'Rapporten aan de regering.'* (Sdu) The Hague, 1992.
- Wetenschappelijke Raad voor het Regeringsbeleid *Eigentijds burgerschap.* (SDU) The Hague, 1992.
- Wetenschappelijke Raad voor het Regeringsbeleid *Duurzame risico's: een blijvend gegeven.* (Sdu) The Hague, 1994.
- Wetering, R.A.P.M. and J.B. Opschoor *De milieugebruiksruimte als uitdaging voor technologie-ontwikkeling.* (Rmno) Rijswijk, 1992.
- Whitebook, J. 'The Problem of Nature in Habermas.' In: *Telos* XD/. (1979) 41-69.
- Whitman, J.P. 'Civil Society and Government: a dispatch from the frontlines.' In: *Public Affairs Quarterly* XV/1 (2001) 17-34.
- Wicksell, K. *Lectures on Political Economy. Volume 1* (Augustus M. Kelly Publishers) London, 1934/1977.
- Wijk, H.D. van, and W. Konijnebelt *Hoofdstukken van administratief recht.* Seventh edition. (Lemma) Utrecht, 1968/1991.
- Wilde, J.W. 'Kosten- en prijsbeheersing in het licht van een dalende aanvoer.' (trans: Cost and Price Management in Light of Decreasing Landings) Paper presented at *Nationale Visserijconferentie*, 1994.
- Williamson, O.E. 'Transaction-Cost Economics: The Governance of Contractual Relations.' In: *Journal of Law and Economics* XXII/. (1979) 233-261.
- Wilson, J.A. 'The Economical Management of Multispecies Fisheries.' In: *Land Economics* DVIII/. (1982) 417-434.
- Wilson, J.A., J.M. Acheson, M. Metcalfe and P. Kleban 'Chaos, Complexity and Community Management of Fisheries.' In: *Marine Policy* XVIII/4 (1994) 295-305.
- Winner, L. *Autonomous Technology: Technics-Out-Of-Control as a Theme in Political Thought.* () Cambridge Mass., 1977.
- Wolin, S.S. 'Revolutionary Action Today.' In: J. Rajchman and C. West (eds.) *Post-Analytic Philosophy.* () New York, 1985.
- World Commission on Environment and Development *Our Common Future.* (Oxford University Press) Oxford, 1987.
- Yeager, P.C. *The Limits of Law. The Public Regulation of Private Pollution.* (Cambridge University Press) Cambridge Mass., 1991.
- Zweers, W. (ed.) *Op zoek naar een ecologische cultuur. Milieufilosofie in de jaren '90.* (Ambo) Baarn, 1991.

INDEX OF NAMES

Issues in Business Ethics

1. G. Enderle, B. Almond and A. Argandoña (eds.): *People in Corporations*. Ethical Responsibilities and Corporate Effectiveness. 1990   ISBN 0-7923-0829-8
2. B. Harvey, H. van Luijk and G. Corbetta (eds.): *Market Morality and Company Size*. 1991   ISBN 0-7923-1342-9
3. J. Mahoney and E. Vallance (eds.): *Business Ethics in a New Europe*. 1992
       ISBN 0-7923-1931-1
4. P.M. Minus (ed.): *The Ethics of Business in a Global Economy*. 1993
       ISBN 0-7923-9334-1
5. T.W. Dunfee and Y. Nagayasu (eds.): *Business Ethics: Japan and the Global Economy*. 1993   ISBN 0-7923-2427-7
6. S. Prakash Sethi: *Multinational Corporations and the Impact of Public Advocacy on Corporate Strategy*. Nestle and the Infant Formula Controversy. 1993
       ISBN 0-7923-9378-3
7. H. von Weltzien Hoivik and A. Føllesdal (eds.): *Ethics and Consultancy: European Perspectives*. 1995   ISBN Hb 0-7923-3377-2; Pb 0-7923-3378-0
8. P. Ulrich and C. Sarasin (eds.): *Facing Public Interest*. The Ethical Challenge to Business Policy and Corporate Communications. 1995
       ISBN 0-7923-3633-X; Pb 0-7923-3634-8
9. H. Lange, A. Löhr and H. Steinmann (eds.): *Working Across Cultures*. Ethical Perspectives for Intercultural Management. 1998   ISBN 0-7923-4700-5
10. M. Kaptein: *Ethics Management*. Auditing and Developing the Ethical Content of Organizations. 1998   ISBN 0-7923-5095-2; Pb 0-7923-5096-0
11. R.F. Duska (ed.): *Education, Leadership and Business Ethics*. Essays on the Work of Clarence Walton. 1998   ISBN 0-7923-5279-3
12. R. Mohon: *Stewardship Ethics in Debt Management*. 1999   ISBN 0-7923-5747-7
13. P.H. Werhane and A.E. Singer: *Business Ethics in Theory and Practice*. Contributions from Asia and New Zealand. 1999   ISBN 0-7923-5849-X
14. R.J. Burke and M.C. Mattis (eds.): *Women on Corporate Boards of Directors*. International Challenges and Opportunities. 2000   ISBN 0-7923-6162-8
15. J.M. Lozano: *Ethics and Organizations*.Understanding Business Ethics as a Learning Process. 2000   ISBN 0-7923-6463-5
16. B. Goodwin: *Ethics at Work*. 2000   ISBN 0-7923-6649-2
17. P. Koslowski: *Principles of Ethical Economy*. 2001   ISBN 0-7923-6713-8
18. W. Dubbink: *Assisting the Invisible Hand*. Contested Relations Between Market, State and Civil Society. 2003   ISBN 1-4020-1444-9

KLUWER ACADEMIC PUBLISHERS – DORDRECHT / BOSTON / LONDON